THE MATH TUTOR'S FRIEND

inequalities

absolute value

binomials

algebraic patterns

divisibility

primes

BOOK 3

ALGEBRA II

by
Joe Cyr

The Math Tutor's Friend *Book 3: Algebra II*

Cyr, Joseph

The Math Tutor's Friend / Book 3 – Algebra II / by Joe Cyr

Third Edition ISBN: 978-1-967039-44-9 Joseph Cyr
Second Edition ISBN: 979-8-9879759-3-0 Joseph Cyr
First Edition ISBN: 9798472688536 Amazon Kindle, 2022

3 4 5 6 7 8 9

Subjects: Main category of book – *absolute values; inequalities; factoring & expanding; quadratics; rational functions; partial fraction expansion* – high-school math

2.Other category - math tutor reference

Cover illustrations by Ramon E. Owen. Cover Design by Joe Cyr

Book design by Joe Cyr

The Math Tutor's Friend
Series

Books

1 - Pre-algebra

2 - Algebra I

3 - Algebra II

4 - Algebra III

5 - Probability & Statistics

6 - Geometry

7 - Trigonometry

8 - Pre-calculus

9 - Series index & "FLASHMATH" handouts/reference cards

visit www.joe-cyr.com

Op. 0546

THE MATH TUTOR'S FRIEND

BOOK 3

ALGEBRA II

Sections

About the "Math Tutor's Friend" Series:

Teachers impart knowledge guided by lesson plans in a course syllabus. Math tutors, on the other hand, face varying topics at differing levels in successive sessions; for example, pre-algebra, advanced algebra, geometry, trigonometry, statistics, or even calculus! This series addresses the issue and is intended to provide the tutor with needed quick review of high-school math topics. The material is not presented rigorously nor does it abound with proofs and explanation that are available in textbooks - the tutor already knows this stuff - but a quick refresher is always a good move before diving into a tutoring session.

Too many in high-school have reached the 11th and 12th grades lacking basic math skills such as knowing the multiplication table (relying on a calculator to multiply 6 by 7, for example), handling signed numbers, and performing operations involving fractions - knowledge that should have been acquired by the fifth grade! I try to address this deficiency by devoting the first few tutoring sessions to these fundamentals and providing handouts of multiplication tables,etc.

Books 1 through 8 of this series include a number of quiz questions (answers explained) that the math tutor may find useful.

Book 9 is an overall index of topics. It contains a set of summary "FLASHMATH" pages for quick reference or for use as handouts.

This third edition corrects typos found in previous editions and expands discussion of some of the topics.

Happy Tutoring!

Joe Cyr

Sarasota, FL

•

THE MATH TUTOR'S FRIEND
|-3|=3
|x|
abs(x)
It's easy !!
|3|=3
(-∞, a]
[a, b)
[b, +∞)
Mr. Math
ABSOLUTE VALUES

About Book 3: "Algebra II"

Section 1, *Absolute Values*. The definition, the absolute-value function, handling absolute-value equations, graphing the absolute-value function.

Emphasis -:

- Definition of absolute-value
- The number line
- Technique for solving absolute-value equations
- Using graphs to solve absolute-value problems.

Section 2, *Inequalities*. Properties, graphing, defining intervals, absolute-value inequalities.

Emphasis -:

- effect of multiplying inequalities by a negative value
- determining the "shaded area" of an inequality graph
- partitioning the number line into intervals reflecting inequalities
- solving inequalities

Section 3, *Factoring and expanding*, introduces important algebraic patterns, the Fundamental Theorem of Arithmetic, and provides shortcuts and handy techniques for factoring functions and using tables to determine the Greatest Common Factor and Least Common Multiple simultaneously.

Emphasis -:

- Divisibility rules
- Factoring integers completely
- Table method for determining common factors
- Distributive Law
- Algebraic patterns

Section 4, *Quadratics,* introduces various methods to factor quadratic functions.

Emphasis -:

- Exactly two solutions, both real or both complex
- factoring quadratics
- what is the depressed quadratic?

Section 5, *Rational functions,* introduces end behavior and the concept of asymptotes and holes.

Emphasis -:

- Determine undefined values of f(x) at the outset

Section 6, *Partial Fraction Expansion,* the opposite of factoring.

Emphasis -:

- The three cases.

• •

1. ABSOLUTE VALUES

TABLE OF CONTENTS

•

1. ABSOLUTE VALUES

Absolute value

Simply stated, the ***absolute value*** of a number x, written as $|x|$, is that number without its sign; it is the distance from 0 on the *number line* and so is always positive or zero. Just think of the absolute value of a number as that number stripped of its sign: $|3| = 3$ and $|-3.7| = 3.7$

Number line and the absolute value |4| and |-4|, showing that 4 is the distance from 0

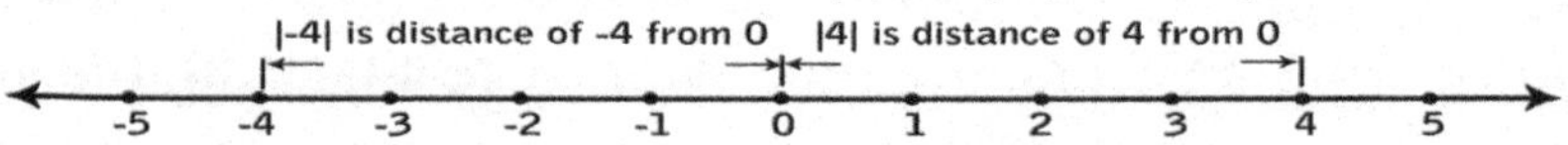

Absolute value function

The *absolute value function* is a function that contains one or more algebraic expression within absolute value symbols "| |". The *absolute value parent function* is $f(x) = |x|$ and is defined as follows:

$$|x| = x \text{ if } x \geq 0 \text{ and } |x| = -x \text{ if } x < 0. \text{ Note that } |x| = 0 \text{ if } x = 0.$$

The absolute value of x frequently appears on calculators and computer apps as "abs(x)." A useful definition of the absolute value is the identity $|x| \equiv \sqrt{x^2}$.
In summary, the absolute value of a number or function is always positive or zero.

Properties of absolute values

(1) $|x| \geq 0$

(2) $|x{\cdot}y| = |x| \cdot |y|$

(3) $|x + y| \leq |x| + |y|$

(4) If $|x| \leq a$ then $-a \leq x \leq a$

(5) If $|x| \geq a$ then $x \leq -a$ or $x \geq a$

(6) $|x - y| < |z - w|$ if and only if $(x - y)^2 < (z - w)^2$

(7) $|x| = \sqrt{x^2}$

Minus signs associated with absolute values

The positioning of exponents and minus signs affects absolute values:

$$-|(-7)^2| = -49$$

Using the *order of operations* rules, first square -7 to get 49 (the square of a negative number is positive). Since $|49| = 49$, then $-|49| = -49$.

$$-(|-7|)^2 = -49$$

In the equation above, the minus sign precedes and is outside of the absolute value sign, so the answer will be negative (because the absolute value of a negative number is always positive).

$$(-|-7|)^2 = 49$$

Although the minus sign precedes and is outside of the absolute value sign, it is <u>inside</u> the parentheses which have a square exponent, and the square is positive.

How to handle absolute-value equations

(a) Place all absolute-value expressions on one side of the equation.

(b) Consider the two possibilities (based on the definition of absolute value):
(1) The expression within "| |" is positive or zero;
(2) The expression within "| |" is negative.

(c) Solve each case for the variable (e.g. x); there will be two answers (depending on whether or not the expression within the absolute-value bars is positive or negative.

(d) Always confirm your answers graphically or by re-inserting the answer into the original equation (or inequality); an answer may be *extraneous* and thus not a solution.

Examples of absolute-value functions:

$|3x - 2| + 4 = 3$ No solution (why?)

$|2x - 5| = 1$ Examine the two possibilities: either ***(a)*** $2x - 5 > 0$ or ***(b)*** $2x - 5 < 0$.

(a) $2x - 5 > 0$: Remove the absolute-value bars and solve for x: $2x = 6$ and $x = 3$.

(b) $2x - 5 < 0$: $-2x + 5 = 1$ $-2x = -4$ $x = 2$

(make one side of the equation the negative of the original; here, change $2x - 5$ to $-2x + 5$). Note: For ***(b)*** we could have used $2x - 5 = -1$, then $2x = -1 + 5$ to get the same result, i.e., $x = 2$.

Thus, $x = 3$ or $x = 2$. Now confirm these values in the original equation, $|2x - 5| = 1$.

$$x = 3: |2 \cdot 3 - 5| = |6 - 5| = |1| = 1 \checkmark$$

$$x = 2: |2 \cdot 2 - 5| = |4 - 5| = |-1| = 1 \checkmark$$

Therefore, both solutions are valid (i.e., neither is extraneous).

If you have two or more absolute-value expressions in an equation, the problem becomes more difficult, for example: $|x - 4| = 3 + |x + 1|$. First, move all absolute-value expressions to one side of the equation (generally on the left):

$$|x - 4| - |x + 1| = 3\,. \qquad (1\text{-}1)$$

By inspection it is obvious* that $x = 0$ is a solution to Equation (1-1) because

$$|0 - 4| - |0 + 1| = |-4| - |1| = 4 - 1 = 3\,.$$

Are there other solutions? Let's find out! Begin with $|x - 4| - |x + 1| = 3$. We shall search for the "*turning points*" (sometimes called "*critical points*" or "*breakpoints*") which are the values of x that make the expression within the absolute-value bars equal to zero. Thus, for $|x - 4|$, simply solve $x - 4 = 0$. The value is $x = 4$, and for $|x + 1|$ the turning-point value is $x = -1$.

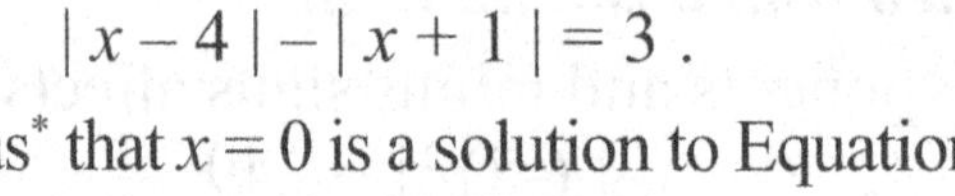

* Always inspect equations before doing algebraic manipulations; you might find a solution "by inspection."

Now mark these values on the number line (bottom of previous page); divide the line into three intervals: (-∞, -1), (-1, 4), and (4, ∞). See the table below.

$\|x-4\|-\|x+1\|=3$			
Interval	(-∞, -1)	(-1, 4)	(4, ∞)
You select a value for x in the interval →	-2	0	5
The sign of $x-4$ is	-	-	+
The sign of $x+1$ is	-	+	+

Recall from the definition of absolute value: $|x| = x$ if $x \geq 0$ and $|x| = -x$ if $x < 0$;:

$$|x-4| = (x-4) \text{ if } x-4 > 0 \text{ and } |x-4| = -(x-4) \text{ if } x-4 < 0.$$

Now, select a value for x in each of the intervals. To keep the computations simple, choose an "easy" value for x in each interval. Now determine which case applies by noting "-" or "+" for the argument of each absolute value when the selected value of x is used, as shown above. Having established the sign (- or +) for the expression within the absolute value bars for each selected value, we apply the definition to determine if the equality is true after solving the equation for x. For this example, select values -2, 0, and 5 in each of the three intervals. Refer to the table above. The resulting pairs of signs: (- -, - +, + +) determined below are shown above the number line.

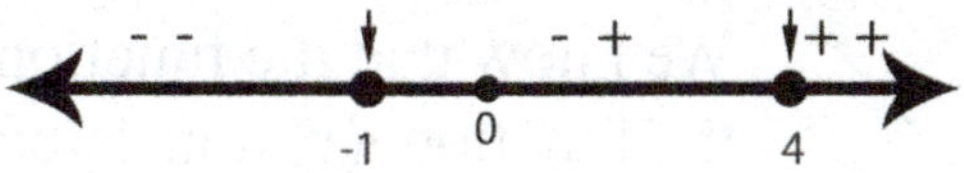

For the first interval, the value $x = -2$ yields a negative argument within both absolute-value bars (*i.e.*, $-2-4=-6$ and $-2+1=-1$), yielding (– –) so according to the definition of absolute value, we take the negative of both arguments and solve Equation (1-1) for x:

$$|x-4|-|x+1|=3. \qquad (1\text{-}1)$$

[Take the negative of terms within bars]: $-(x-4)-[-(x+1)] \stackrel{?}{=} 3$

[Remove the parentheses and bracket]: $-x+4+x+1 \stackrel{?}{=} 3$ ***[False]***

Here, the variable x cancels out and we get $4+1=3$, which clearly is not true, so no value of x in the interval (-∞, -1) will satisfy the equation $|x-4|-|x+1|=3$.

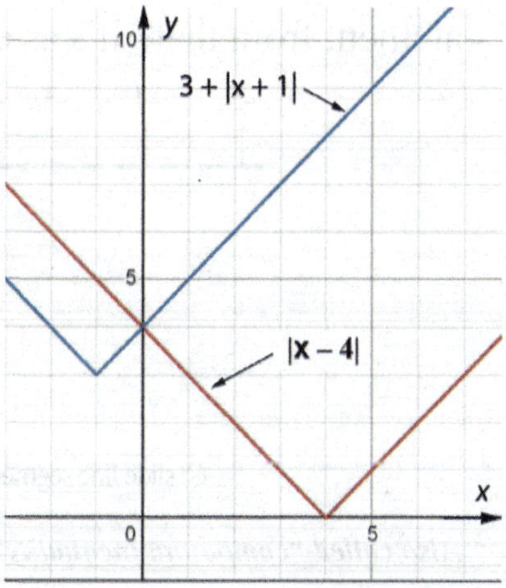

Consider the interval (-1, 4) and choose $x = 0$ in that interval, which gives a negative value for the term $(x-4)$ and a positive value for the term $(x+1)$, yielding (– +) so we take the negative of the first term and make no change to the second term:

$$-(0-4)-(0+1) \stackrel{?}{=} 3$$

$$4-1=3 \checkmark$$

and the resulting $x = 0$ is a correct solution.

Finally, check the third interval (4, ∞); choose 5.

At $x = 5$, both $x - 4$ and $x + 1$ are > 0, yielding (+ +), so, just remove the absolute-value bars:

$$|x - 4| - |x + 1| \stackrel{?}{=} 3$$

$$x - 4 - x - 1 = 3 \quad \textit{[False]}$$

The xs cancel yielding -5 = 3, which clearly is not true. Therefore, $x = 0$ is the only solution to $|x - 4| - |x + 1| = 3$. See the graph on the previous page.

Summarizing, there are two absolute-value terms in the function under analysis, $|x - 4|$ and $|x + 1|$. For the test value of x used in each of the intervals, look at the sign of the term within the absolute-value bars, i.e., $(x - 4)$ and $(x + 1)$. For the first interval we selected $x = -2$. This selection causes "$x - 4$" in $|x - 4|$ to be -6, a negative number, thus by definition of the absolute value, we have $-(x - 4)$; that is, because $x - 4$ is negative we take its negative to get a positive number: $-(-2 - 4) = -(-6) = 6$. That's why we determined the sign of each absolute-value term in the function; if the sign is negative, we take the negative of the term within the absolute-value bars (| |); if the sign is positive we make no changes.

Graphing the absolute-value function

Graph the parent function $y = |x|$ and the translated function $|x - 3| - 4$. Informally, the graph of an absolute value *linear* function is always a "V," as seen to the left (and in the previous example).

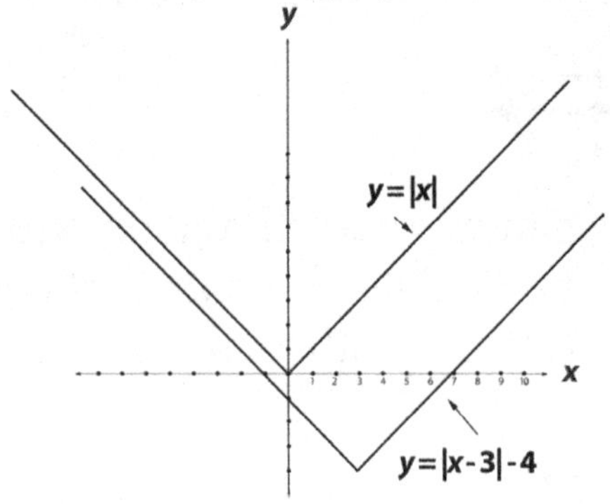

We know that the function $f(x - h) + k$ is simply the function $f(x)$ translated to the right along the x-axis by h units, and up along the y-axis by k units (down, if $k < 0$). Therefore, if we are given the absolute value function

$$f(x) = y = |x - 3| - 4$$

the *parent function* "V" moves to the right by 3 units and down by 4 units. The solution (x-intercept) is $x = -1$ and $x = 7$ (check by substituting each value in the function).

Absolute value inequality statement from a double inequality

Consider the double inequality* $-2 < x < 4$. How does one make an absolute-value statement from that? It's as easy as ABC: look at the number line shown below!

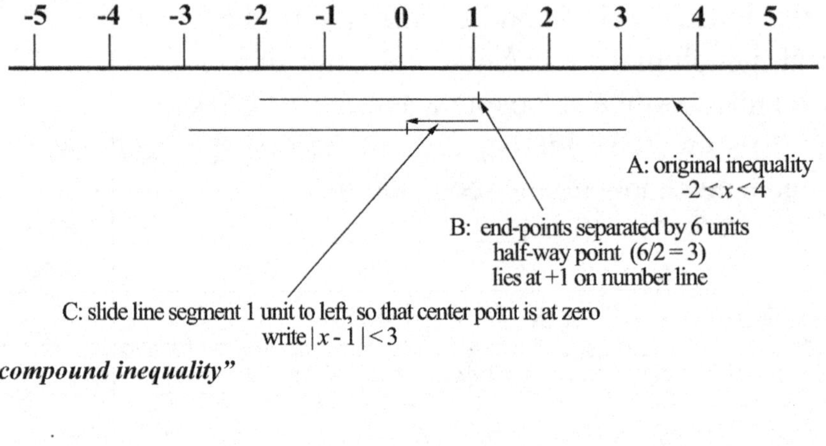

* *Also called "compound inequality"*

Or, manipulate the double inequality to get the same absolute value on either side (add -1 to all sides): $-3 < x - 1 < 3$ or $|x - 1| < 3$.

Using graphs to solve absolute-value equations

Example: Solve $|2x - 3| - 4 = 6$.

Simply draw a graph for each side of the equation. The solutions are $x = -3.5$ and $x = 6.5$. Verify this by solving the equation by hand. Using graphs can help solve absolute-value problems quickly.

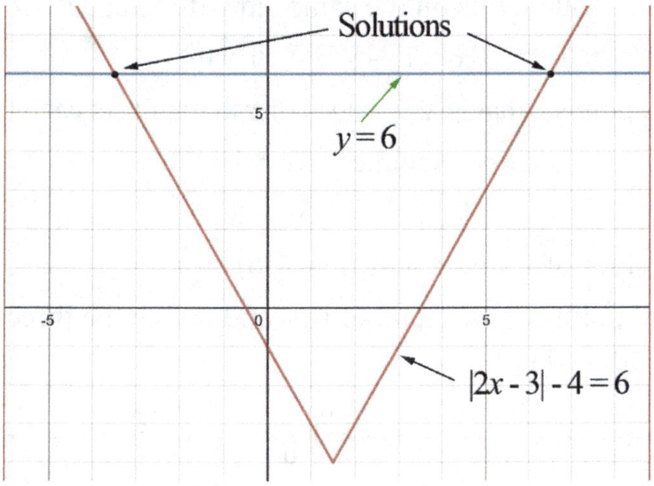

Absolute value problem

Example: $|3x + 1| = |-5x + 8|$

We'll look at three ways to approach the solution to this problem"

(1) Direct solution; (2) "Critical point" solution; and (3) Graphical solution.

Direct Solution

Example: $|3x + 1| = |-5x + 8|$

Apply the definition:

$|x| = x$ for $x \geq 0$ (just remove the bars), and $|x| = -x$ for $x < 0$ (take the negative of what's inside the bars)

(assume $3x + 1 \geq 0$): $3x + 1 = |-5x + 8|$ (1-2)

(assume $3x + 1 < 0$): $-(3x + 1) = |-5x + 8|$ (1-3)

Now rewrite Equations (1-2) and (1-3) and apply the definition again

$$|-5x + 8| = 3x + 1$$

(assume $-5x + 8 \geq 0$) [remove bars}: $-5x + 8 = 3x + 1$ (1-2a)

(assume $-5x + 8 < 0$) [negative of term in bars]: $-(-5x + 8) = 3x + 1$ (1-2b)

Solve Equation (1-2a) for x:

$-8x = -7$ [add $-3x$ to both sides; add -8 to both sides of (1.2a)]

$x = 7/8 = 0.875$ [solve for x]

Solve Equation (1-2b) for x:

$5x - 8 = 3x + 1$ [multiply left side of (1-2b) by -1]

$2x = 9$ [add $-3x$ to both sides; add 8 to both sides]

$x = 4.5$ [solve for x] .

Verify the two solutions 0.875 and 4.5 by substituting for x in the original equation

$$|\,3x + 1\,| = |\,-5x + 8\,|$$

$|\,3\times0.875 + 1\,| = |\,-5\times0.875 + 8\,|$? $3.625 = 3.625$ ✓ ***check!***

$|\,3\times4.5 + 1\,| = |\,-5\times4.5 + 8\,|$? $|\,14.5\,| = |\,-14.5\,|$ ✓ ***check!***

You will get the same results if you solve (1-3) for x. Try it!

Critical Point Solution

<u>Example</u>: $|\,3x + 1\,| = |\,-5x + 8\,|$

Put both absolute-value terms on the same side of the equation

$$|3x + 1| - |-5x + 8| = 0$$

Find the "critical points;" that is, the values of x that will set each term to zero:

$$x = -\tfrac{1}{3} \text{ makes } |\,3x + 1\,| = 0$$

$$x = \tfrac{8}{5} \text{ makes } |\,-5x + 8\,| = 0$$

so the critical values of x are $-\frac{1}{3}$ and $\frac{8}{5}$ (= 1.6).

Mark the critical points on the number line and define the three intervals $(-\infty, -\frac{1}{3})$, $(-\frac{1}{3}, 1.6)$, and $(1.6, \infty)$.

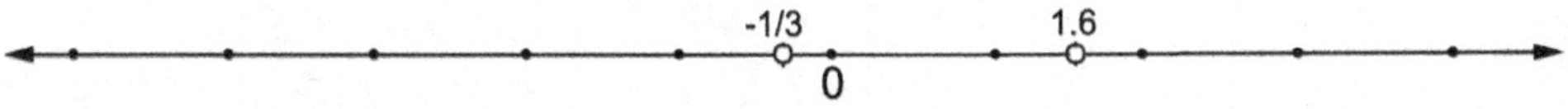

The contents of each absolute-value term will be either positive (" + ") or negative (" - ") in each interval.

Set up a table as shown below, then choose a value for x in each interval (choose and "easy" value to simplify calculation), then determine the sign ("*sgn*") of the contents of each term in each interval when using the selected value of x:

$|\,3x + 1\,| - |\,-5x + 8\,| - 0$			
Interval	$(-\infty, -\frac{1}{3})$	$(-\frac{1}{3}, \frac{8}{5})$	$(\frac{8}{5}, \infty)$
Choose x →	-1	0	2
$sgn(3x + 1)$	-	+	+
$sgn(-5x + 8)$	+	+	-

Now ignore the absolute-value symbol and solve the equation in each interval using the sign determined above for each absolute-value term:

Interval $(-\infty, -\frac{1}{3})$: $-(3x + 1) - (-5x + 8) = 0$

Solving for x: $x = 4.5$.

Interval $(-\frac{1}{3}, \frac{8}{5})$: $(3x + 1) - (-5x + 8) = 0$

Solving for x: $x = \frac{7}{8} = 0.875$.

Interval $(\frac{8}{5}, \infty)$: $(3x + 1) - [-(-5x + 8)] = 0$

Solving for x: $x = 4.5$.

The two solutions for x are 0.875 and 4.5 which we already determined and verified.

Graphical Solution

Example: $|3x + 1| = |-5x + 8|$

Graph the expressions on both sides of the "=" sign. As seen in the figure on the right, the solutions are at the intersections .

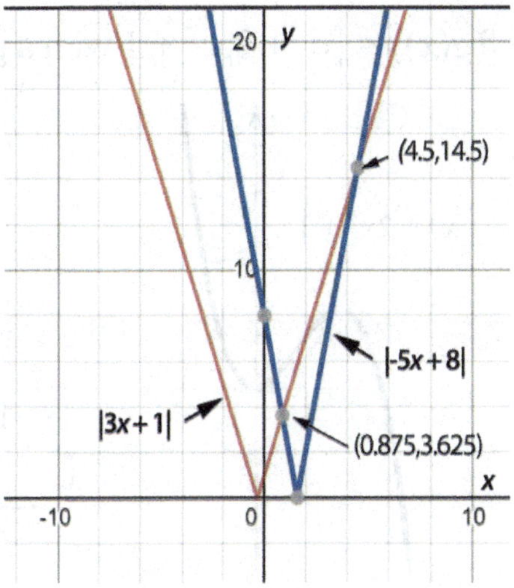

Example: $|x^3 - 2| - |2x^2| - x^2| = x$

Graph the expressions on both sides of the "=" sign. As seen below, the four solutions are at the intersections of the expressions on both sides of the equation.

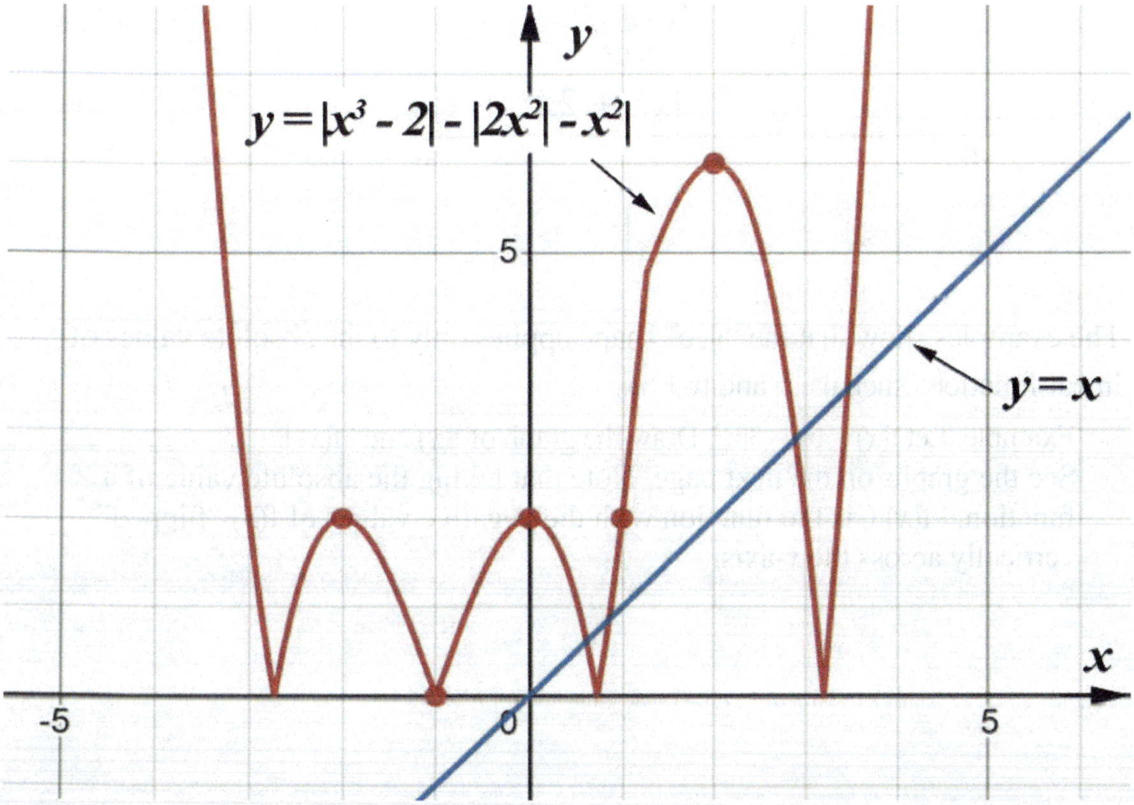

Composite absolute-value functions

If $f(x) = x^3 + 2x^2 + 1$ and $g(x) = |x|$ then the composite function

$f(g(x)) = |x|^3 + 2|x|^2 + 1$ with a graph as shown below.

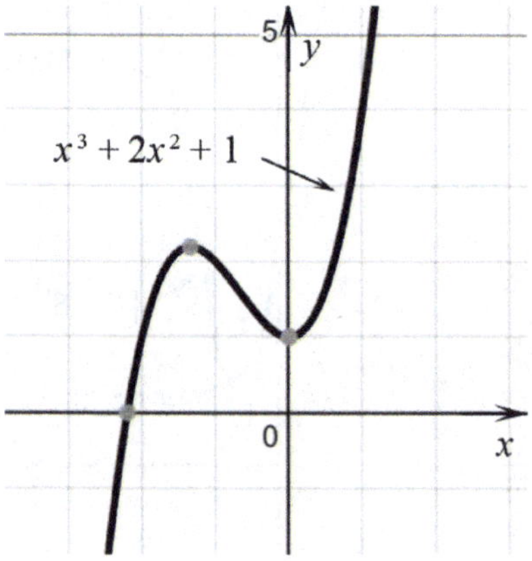

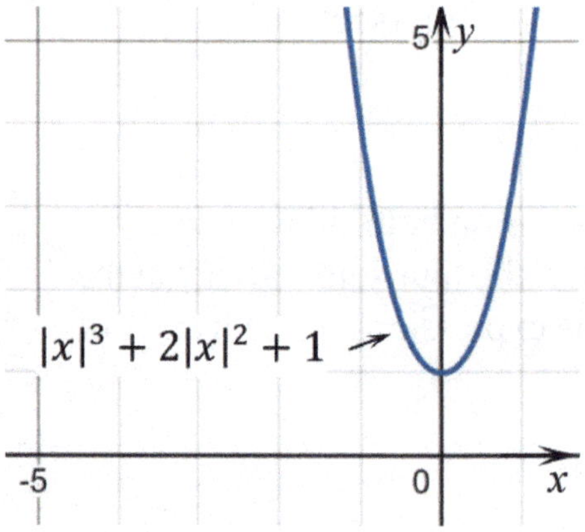

What about g(f(x)), i.e., $|f(x)| = |x^3 + 2x^2 + 1|$? Here's the graph of g(f(x)) :

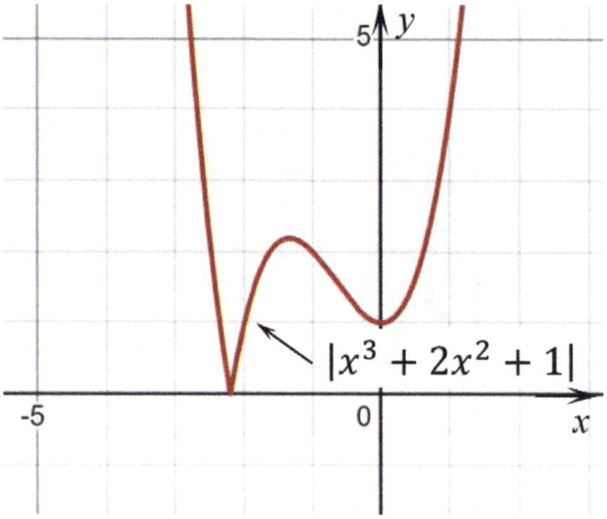

The examples show that the "vee" shape applies only to the absolute values of linear functions such as $|x|$ and $|x + a|$.

<u>Example</u>: Let $f(x) = |x|^3 - 3|x|^2$. Draw the graph of f(x) and | f(x) |.
See the graphs on the next page. Note that taking the absolute value of a function, | f(x) |, is the function with the negative values of f(x) "flipped" vertically across the x-axis.

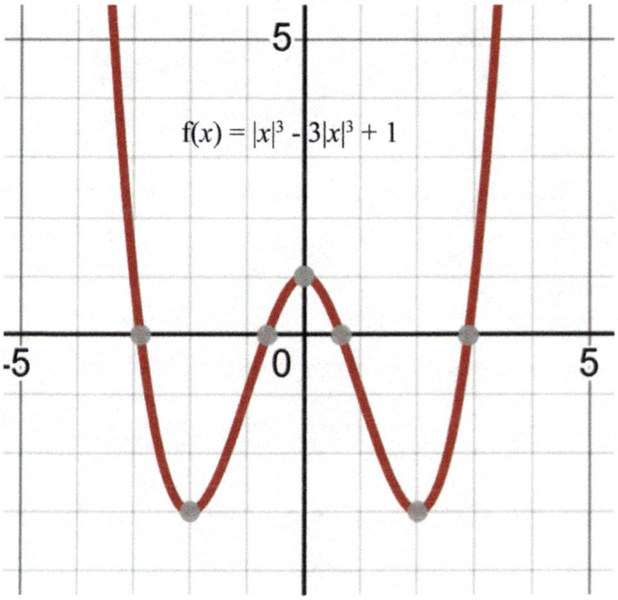
5
f(x) = |x|³ - 3|x|³ + 1
-5
0
5

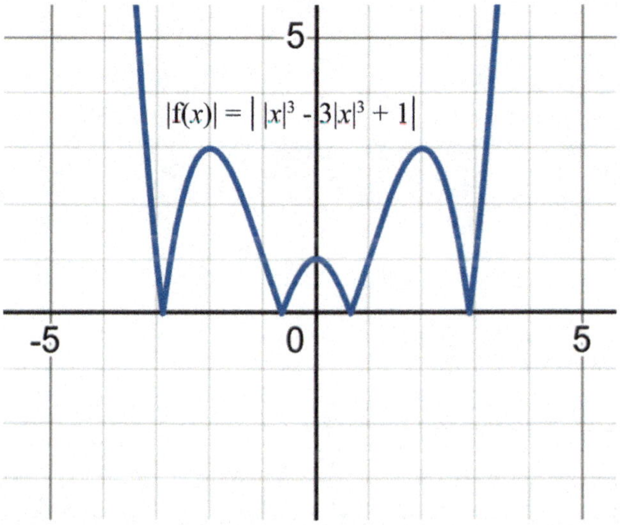
5
|f(x)| = | |x|³ - 3|x|³ + 1|
-5
0
5

•

THE MATH TUTOR'S FRIEND
STRICT INEQUALITY
$3 \geq x \geq -5$
$x^2 - 2 > 0$
[a,b)
It's easy !!
number line
$\frac{(x-2)^2(x+3)}{4-x} < 0$
$X - Y < 0$
Mr. Math
IN-EQUALITIES

2. INEQUALITIES

TABLE OF CONTENTS

2. INEQUALITIES

Inequality symbols

When comparing two numbers, clearly they may be equal "$=$" ($4=4$) or not equal "$\neq$" at all ($5 \neq 6$) where one number may be larger ($6 > 5$) or smaller ($5 < 6$) than the other. This is also true for variables and expressions; that is, one may be equal to ($a = b$) or larger ($x > y$) or smaller $(x+1) < (z^2 - 4)$ than the other.

The symbols $<$ and $>$ are *inequality symbols*. Further, these two symbols represent *strict inequalities*. Here are the *weak inequality* (also *non-strict inequality*) symbols: $\leq$ and $\geq$ which are generally read as "less than or equal to" and "greater than or equal to," respectively. Because most computer keyboards have the strict inequality symbols but not the weak inequality symbols, the following substitutions are sometimes used: " <= " and " >= ". The composite symbol " <> " is sometimes used to represent "$\neq$". However, many word-processing programs allow the insertion of special symbols, and so <Alt> 242 and <Alt> 243 produce* "$\geq$" and "$\leq$." Rules for handling *equations* in mathematics can be extended to include *inequalities* with a few additions. Here are some other symbols relating to inequality: " << " means *much less than*, and " >> " means *much greater than*. If an inequality symbol is changed, say "$<$" changes to "$>$", then the *sense* of the inequality is changed (the term "*invert*" is sometimes used instead of "sense"). Finally, there is a *chained notation* (also called *compound inequality*) "$0 < a \leq b$," that means a is both greater than zero (i.e., positive) **and** less than or equal to b." Clearly, if $a \leq b$ then $b \geq a$. Note that the "arrow" of the inequality symbol always points to the smaller value.

Properties of inequalities

The following properties also hold if the weak inequalities ($\leq$ and $\geq$) shown are replaced by their corresponding strict inequalities ($<$ and $>$).

Transitivity: For any real numbers a, b, c:

- If $a \geq b$ and $b \geq c$, then $a \geq c$.
- If $a \leq b$ and $b \leq c$, then $a \leq c$.

Converse: For any real numbers a and b:

- If $a \leq b$, then $b \geq a$.
- If $a \geq b$, then $b \leq a$.

Addition and Subtraction: A constant c may be added to or subtracted from both sides of an inequality:
For any real numbers a, b, c

- If $a \leq b$, then $a + c \leq b + c$ and $a - c \leq b - c$.
- If $a \geq b$, then $a + c \geq b + c$ and $a - c \geq b - c$.

* Hold down the <Alt> key and use the numerical keypad; make sure that <Num Lock> is on.

Multiplication and division: If $x < y$ and $a > 0$ (*i.e.*, a is positive), then $ax < ay$.

If $x < y$ and $a < 0$ (*i.e.*, a is negative), then $ax > ay$.

[Note the change in sense of the inequality]

For any real numbers, a, b and non-zero c:

If c is positive, then multiplying or dividing by c does not change the inequality:

- If $a \geq b$ and $c > 0$, then $ac \geq bc$ and $a/c \geq b/c$.
- If $a \leq b$ and $c > 0$, then $ac \leq bc$ and $a/c \leq b/c$.

If c is negative, however, then multiplying or dividing by c inverts the inequality:

- If $a \geq b$ and $c < 0$, then $ac \leq bc$ and $a/c \leq b/c$.
- If $a \leq b$ and $c < 0$, then $ac \geq bc$ and $a/c \geq b/c$.

[Note the change in sense of the inequality]

Additive inverse: For any real numbers a and b, negation (multiplying by -1) inverts (changes the sense of) the inequality:

- If $a \leq b$, then $-a \geq -b$
- If $a \geq b$, then $-a \leq -b$.

Multiplicative inverse: For any non-zero real numbers a and b that both are positive or both negative:

- If $a \leq b$, then $1/a \geq 1/b$.
- If $a \geq b$, then $1/a \leq 1/b$.

If one of a and b is positive and the other is negative, then:

- If $a < b$, then $1/a < 1/b$.
- If $a > b$, then $1/a > 1/b$.

The important thing to remember about inequalities is that if an inequality is multiplied or divided by a negative number, then the sense of the inequality changes (*i.e.*, is inverted).

Exponents and logarithms: Raising both sides of an inequality to a power $n > 0$ when a and b are positive real numbers:

- If $a \leq b$, then $a^n \leq b^n$.
- If $a \leq b$, then $a^{-n} \geq b^{-n}$.

Taking the natural logarithm to both sides of an inequality when a and b are positive real numbers:

- If $a \leq b$, then $\ln(a) \leq \ln(b)$. This is also true for logarithms to any base.

Graphing inequalities

The graph (top of next page) on the left is that of the function $y = (x - 1)^2 - 1$; its parent function ($y = x^2$) is translated one unit to the right and one unit down. Consider the inequality $y > (x - 1)^2 - 1$ which states that y is above the line and does not include points on the dashed line, as illustrated by the shaded area shown in the graph on the right of the figure.

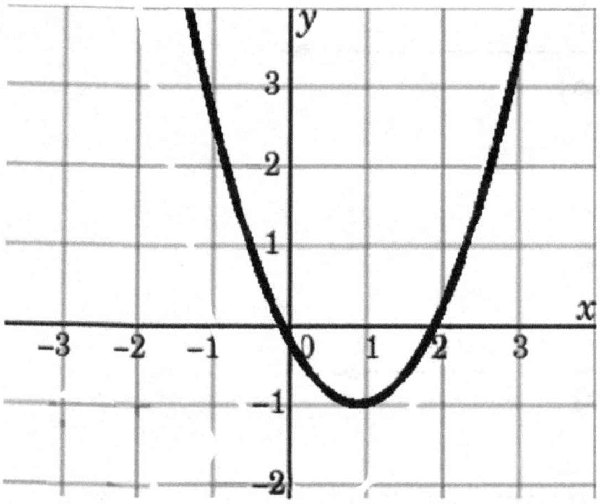

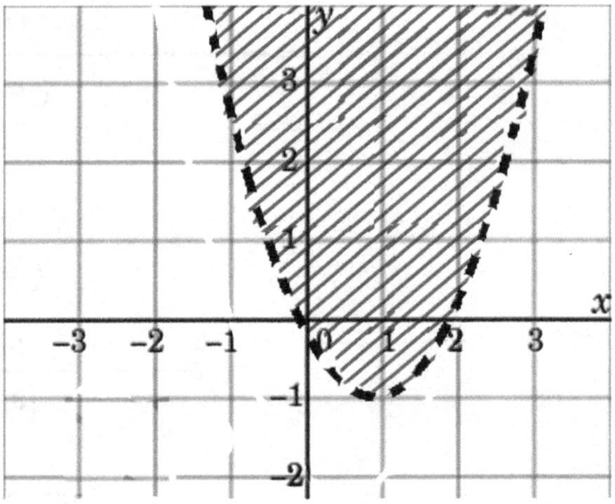

Now, consider the weak inequality $y \geq (x - 1)^2 - 1$. Such inequalities are shown by a solid line on the graph, which means that y may have values in the shaded area, and also take on the values on the graph line. These features can be reproduced on a graphing calculator which can shade an area above (or below) a graphed inequality function.

Examples of inequality functions

Multiplying or dividing an inequality by (-1) changes the *sense* of the inequality.

Example: Multiply $y \geq (x - 1)^2 - 1$ by negative one:

$$(-1)\,y \leq (-1)[(x - 1)^2 - 1] \qquad \text{[Inequality is “flipped”]}$$

$$-y \leq -[(x - 1)^2 - 1]$$

Now add $[(x - 1)^2 - 1] + y$ to both sides to get

$$[(x - 1)^2 - 1] + y - y \leq -[(x - 1)^2 - 1] + [(x - 1)^2 - 1] + y$$

$$[(x - 1)^2 - 1] \leq y$$

which is the original inequality (strictly speaking, it is the converse of the original inequality).

A word about intervals

An interval between values a and b (from a to b) on the number line that includes both a and b is denoted by the *closed interval* $[a,b]$; if it includes a but not b, the notation is the *half-open interval* (or *half-closed interval*) $[a,b)$; similarly for $(a,b]$. If neither a nor b are included in the interval, the *open interval* notation (a,b) is used. On the number line, inclusion is denoted by a solid line; non-inclusion is included by dashes or open dots. If infinity is part of an interval, it clearly cannot be included as it is not defined, so the notation is (∞, ∞) or $(-\infty, a]$; $(-\infty, a)$; or (b, ∞); or $[b, \infty)$. See the figure at the top of the following page.

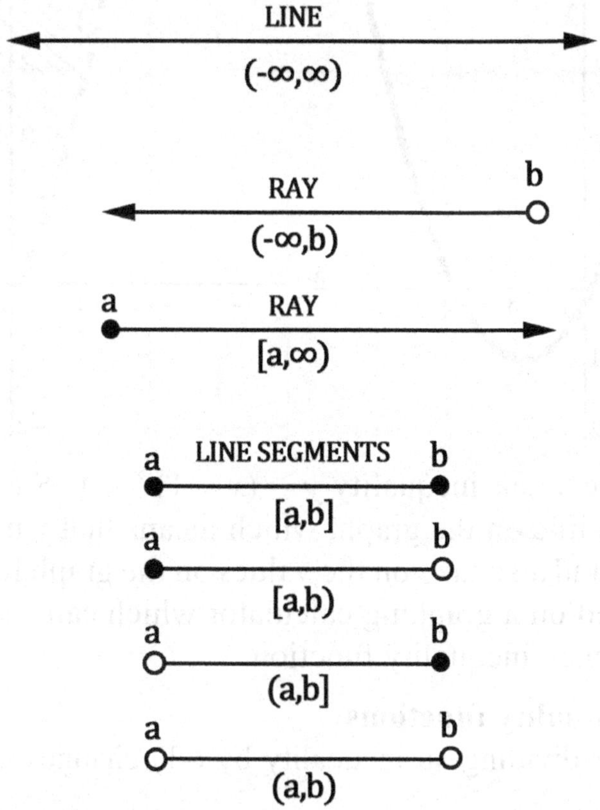

Rational inequalities

A rational function is a fraction with a polynomial in the numerator and a polynomial in the denominator. It may be expressed as $f(x) = P(x)/Q(x)$. $f(x)$ is defined only at values of x such that $Q(x) \neq 0$. Just as an integer is a rational number having a 1 in the denominator, so too is $f(x) = P(x)/1$ a rational function (in such a case, $Q(x) = 1$). Another section of this book is devoted to rational functions.

Solving a rational inequality

1) If necessary rewrite the inequality so that one side is zero.
2) Find all of the linear factors of $P(x)$ and $Q(x)$.
3) Find the critical values of each linear factor by equating each to zero and solving for the variable (in this case, x). It will be helpful to mark each value on a number scale.
4) Determine the sign of the original function in each interval formed by the critical values.
5) The solution will be those intervals within which the variable allows $f(x)$ to satisfy the inequality.

Examples of rational inequalities

Example 1: $x^3 - x^2 \leq 6x$ [here, $Q(x) = 1$]

First, we rearrange the inequality to make the right side zero by adding $-6x$ to each side:

$$f(x) = x^3 - x^2 - 6x \leq 0$$

which can be factored as follows:

$x(x^2 - x - 6)$ [factor out x]

$= x(x + 2)(x - 3)$ [factor the quadratic]

Now by setting each factor to 0, calculate the three values of x that will make $f(x) = 0$. The values are $x = -2$; $x = 0$; and $x = 3$.

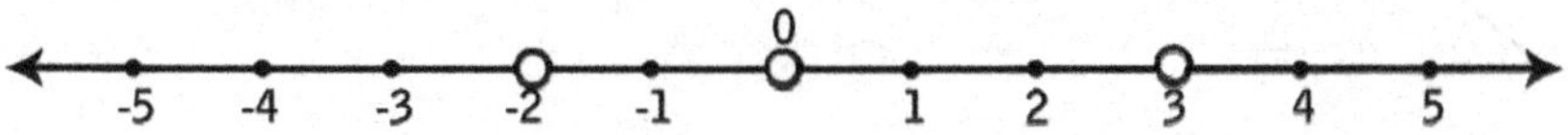

The number line (above) displays four open intervals: $(-\infty, -2)$; $(-2, 0)$; $(0, 3)$; and $(3, \infty)$. Next, choose a value in each interval and determine the sign of $x^3 - x^2 - 6x$. It's wise to select "easy" values.

What does all of this mean? Look at the graph of $y \geq f(x) = x^3 - x^2 - 6x$ shown below. Note that $f(x)$ is negative in the intervals $(-\infty, -2)$ and $(0, 3)$. Therefore, if you want to avoid the interval calculations, simply graph the function and locate them easily, as shown.

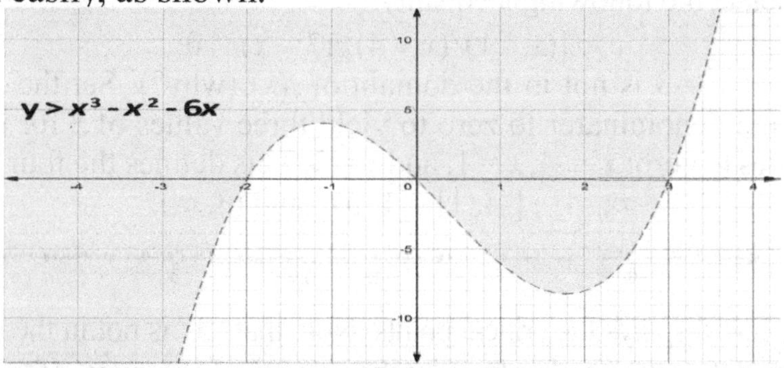

The sign of the result is the product of the signed numbers of each factor; if there are an odd number of negative signs, then the result is negative; otherwise it is positive. The two intervals where the result is negative are $(-\infty, -2)$ and $(0, 3)$. It is always possible to calculate the selected value in each interval using the entire function $f(x)$.

Interval	*Selected Value*	*x*	*(x + 2)*	*(x – 3)*	*Sign of f(x)*
$(-\infty, -2)$	-3	$-3 < 0$	$-1 < 0$	$-6 < 0$	-
$(-2, 0)$	-1	$-1 < 0$	$1 > 0$	$-4 < 0$	+
$(0, 3)$	1	$1 > 0$	$3 > 0$	$-2 < 0$	-
$(3, \infty)$	4	$4 > 0$	$6 > 0$	$1 > 0$	+

Example 2*:*

$$y \leq f(x) = [(x-1)^2(x+4)]/(3-x)$$

The region that satisfies this inequality is shaded in the figure below.

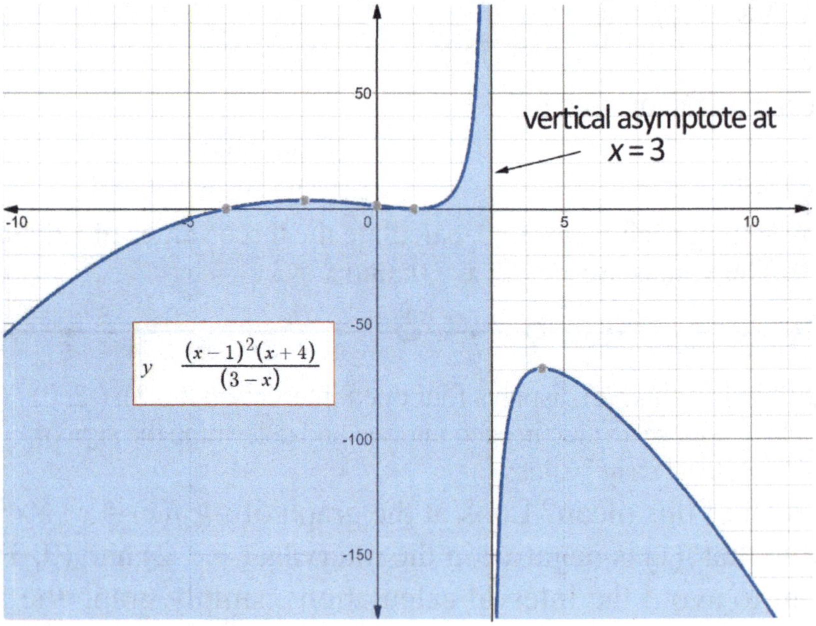

Now let's look at the following inequality:

$$y = [(x-1)^2(x+4)]/(3-x) < 0$$

Note first that $x = 3$ is not in the domain of $f(x)$ (why?). Set the factors of numerator and denominator to zero to yield three values of x for which the term evaluates to zero: $x = -4$, $x = 1$, and $x = 3$. This defines the four intervals:

$$(-\infty, -4]\ ;\ [-4, 1]\ ;\ [1, 3)\ ;\ \text{and } (3, \infty).$$

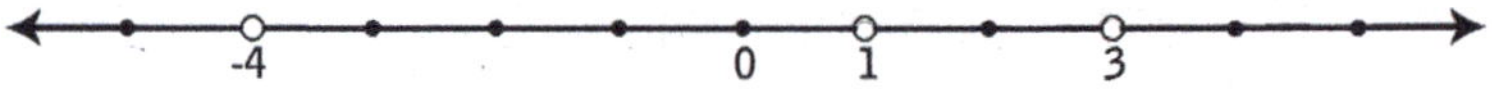

The number line displays the critical points. Note that "3" is not in the domain of $f(x)$. The open dot on the number line signifies a *vertical asymptote* of the function. Construct a table as before. The result is that $f(x)$ is less than zero in two of the intervals: $(-\infty, -4)$ and $(3, \infty)$. The domain of $f(x)$ all x except $x = 3$ (the vertical asymptote).

Note that the function is negative in the intervals $(-\infty, -4)$ and $(3, \infty)$ as determined in the table below, and that there is a vertical asymptote at $x = 3$.

Interval	Selected Value	$(x-1)^2$	$(x+4)$	$(3-x)$	Sign of f(x)
$(-\infty, -4)$	-5	36	-1	8	-
$(-4, 1)$	0	1	4	3	+
$(1, 3)$	2	1	6	1	+
$(3, \infty)$	5	4	9	-2	-

Inequalities involving absolute values

The absolute-value inequality $|\,\boldsymbol{x}\,| < \boldsymbol{a}$ can be restated as

$$|\,\boldsymbol{x}\,| < a \text{ if } x > 0 \text{ and } |\,\boldsymbol{x}\,| > -a \text{ if } x < 0$$

This is from the definition of absolute value $|\,\boldsymbol{x}\,| = \boldsymbol{a}$:

$$|\,\boldsymbol{x}\,| = a \text{ if } x \geq 0 \; |\,\boldsymbol{x}\,| = -a \text{ if } x < 0$$

that is, $$-a < x < a.$$

<u>Example</u>: Consider the absolute-value inequality:

$$|\,x + 3\,| < 5.$$

We convert it to a compound inequality by taking the negative of the right side and placing it to the left as follows to create the *chained notation*:

$$-5 < |\,x + 3\,| < 5$$

Next, remove the bars ("| |")

$$-5 < x + 3 < 5$$

then solve

$$-5 - 3 < x + 3 - 3 < 5 - 3 \qquad \text{[add (-3)]}$$
$$-8 < x < 2 \qquad \text{[simplify]}$$

See the number line below.

Compound inequalities as absolute-value inequalities

This is the reverse of the process discussed in the previous section. Let's develop the general case first and follow with a couple of examples. We wish to convert the compound inequality $a < x < b$ to an inequality of the form $|\,x - n\,| < m$ where n and m are functions of a and b. Now a and b are points on the number line.

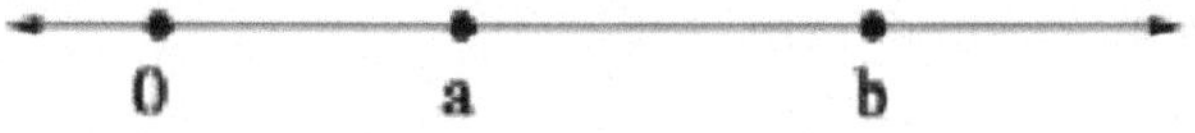

What we are really trying to do is to translate the line segment ab by k units so that it "straddles" the *0* point, providing equal distances between a and zero, and zero and b. That is, $|\,a - k\,| = |\,b - k\,|$. What is this distance k ? It is simply the arithmetical average (mean) of the distance between a and b. So, now we know that $k = (a + b)/2$. Moving the interval (the distance between a and b) by k units

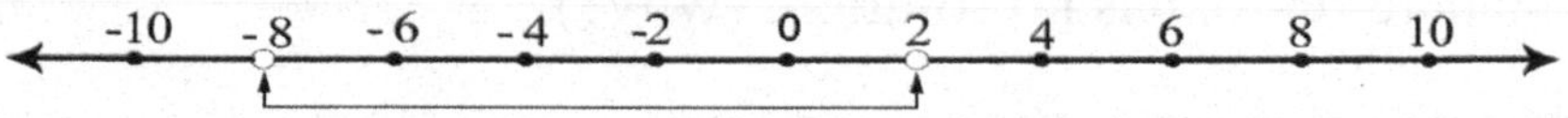

is done by subtracting k from each part of the compound inequality $a < x < b$:

$$a - k < x - k < b - k$$

Since $k = (a + b)/2$, we have $\quad a - (a + b)/2 < x - (a + b)/2 < b - (a + b)/2$
which simplifies to

$$(a - b)/2 < x - (a + b)/2 < (b - a)/2$$

or equivalently (multiply all sides by 2),

$$-(b - a) < 2x - (a + b) < (b - a)$$

yielding

$$|\,2x - (a + b)\,| < (b - a).$$

In summary, to convert the compound inequality

$$a < x < b$$

to an absolute-value inequality, we write

$$|\, 2x - (a + b) \,| < (b - a).$$

Examples:

(a) Convert the compound inequality $1 < x < 5$ to an absolute-value inequality.

Solution:

$1 < x < 5$ becomes $|\, 2x - (1 + 5) \,| < (5 - 1)$, or $|\, 2x - 6 \,| < 4$, that simplifies to $|\, x - 3 \,| < 2$. We simply moved the interval $b - a = 1 - 5$ to the left by $(5 + 1)/2 = 3$ units, so that the compound inequality $1 < x < 5$ becomes $1 - 3 < x - 3 < 5 - 3$ or $-2 < x - 3 < 2$ which is equivalent to $|\, x - 3 \,| < 2$. It's as simple as that!

(b) Convert the compound inequality $-5 < x < 3$ to an absolute-value inequality.

Solution:

Apply the formula directly:

$$|\, 2x - (a + b) \,| < (b - a)$$
$$|\, 2x - (-5 + 3) \,| < 3 - (-5)$$
$$|\, 2x + 5 - 3 \,| < 3 + 5$$
$$|\, 2x + 2 \,| < 8$$
$$|\, x + 1 \,| < 4$$

What if the compound inequality is of the form $a > x > b$? Simply "turn it around" to get it into the desired form:

$$b < x < a.$$

Special cases involving absolute-value inequality

1) Any absolute value inequality of the form

$$|\, x \pm a \,| < b$$

where $b < 0$, has no solution. (why?)

2) An absolute value inequality of the form

$$|\, x \pm a \,| > -b,$$

where $b > 0$, has solution of all x, because any absolute value, being positive or zero, is greater than a negative number.

•

Under what conditions is $|\boldsymbol{x} \pm \boldsymbol{a}| < \boldsymbol{b}$?

The inequality $|x \pm a| < b$ is true under the following conditions:

- $b > 0$: The absolute value represents a distance, which is always non-negative.
- The expression inside the absolute value, ($x \pm a$), is between -b and b:

  The inequality $|\boldsymbol{x} \pm \boldsymbol{a}| < b$ means that the distance of $(\boldsymbol{x} \pm \boldsymbol{a})$ from zero is less than b. This translates to the compound inequality:

  $$-b < \boldsymbol{x} \pm \boldsymbol{a} < b$$

  Solve for x in this inequality**:** There are two cases.

  Case 1: $|x + a| < b$;

  the compound inequality is $-b < x + a < b$, then

  $-b - a < x < b - a$ [subtract 'a' from all parts of the inequality $-b < x + a < b$]

  Case 2: $|x - a| < b$;

  the compound inequality is $-b < x - a < b$, then

  $a - b < x < a + b$ [add 'a' from all parts of the inequality $-b < x - a < b$]

Solve the compound inequalities:

This gives you the range of x values that satisfy the original inequality. Clearly, if $a = 0$ then the compound inequality is simply $-b < x < b$ and the domain of x is $(-b, b)$.

For $a > 0$, the domain partitions into two "mirror" sub-domains reflected across the y-axis.

Let's look at a as a function of b, say, $a = k \times b$ where $k \geq 0$ (k need not be an integer). The table below shows the effect of $k = 0$, 1, 2, and 3 for the case of $b = 2$.

k	$a = k \times b = k \times 2$	domain of $\lvert x \pm a \rvert < b$ for $b = 2$
0	$a = 0 \times 2 = 0$	x ε $(-b, b) = (-2, 2)$
1	$a = 1 \times 2 = 2$	x ε $(-4, 0)$ U $(0, -4)$
2	$a = 2 \times 2 = 4$	x ε $(-6, -2)$ U $(2, 6)$
3	$a = 3 \times 2 = 6$	x ε $(-8, -2)$ U $(2, 8)$
k	$a - k \times 2$	x ε $(-(a + b), -(a - b))$ U $((a - b), (a + b))$

The graph of these inequalities where $b = 2$, $a = 2$, and $b = 4$ is shown below.

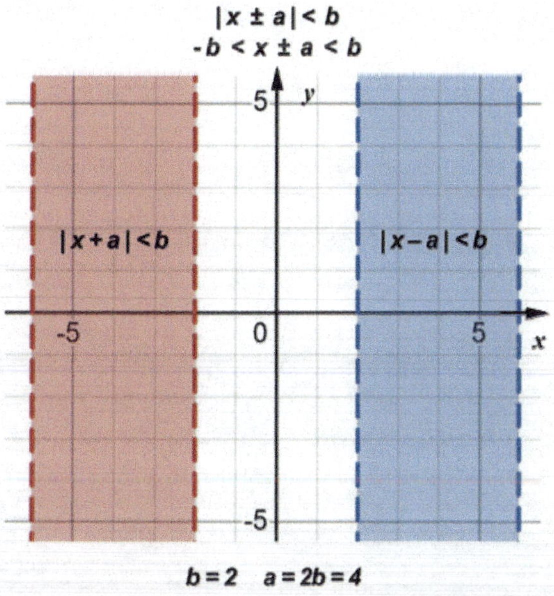

THE MATH TUTOR'S FRIEND
BINOMIAL THEOREM
SYNTHETIC DIVISION
It's easy !!
ZEROES
PASCAL'S TRIANGLE
PRIME
Mr. Math
FACTORING & EXPANDING

3. FACTORING & EXPANDING

TABLE OF CONTENTS

•

3. FACTORING & EXPANDING

Factors and primes

A *factor* of a given number is any number that divides exactly into that number, that is, with a remainder of zero. For example, 3 and 5 are factors of 15*. Another way of stating this is that 15 is *divisible* by 3 and 5. All of the factors of 15 are 1, 3, 5, and 15 itself.

A *prime number*, or *prime*, is a *natural number*** greater than 1 that has exactly two factors: the number itself and 1. That is, a prime number can be divided only by itself and 1. Although zero is divisible by 1 ($0/1 = 0$) it is not a prime number because it cannot be divided by itself: *i.e.*, 0/0 is undefined. The number 1 is not considered to be prime by definition. The first few primes are 2, 3, 5, 7, 11, ...

Here are the prime numbers less than 500:

2, 3, 5, 7, 11, 13, 17, 19, 23, 29, 31, 37, 41, 43, 47, 53, 59, 61, 67, 71, 73, 79, 83, 89, 97, 101, 103, 107, 109, 113, 127, 131, 137, 139, 149, 151, 157, 163, 167, 173, 179, 181, 191, 193, 197, 199, 211, 223, 227, 229, 233, 239, 241, 251, 257, 263, 269, 271, 277, 281, 283, 293, 307, 311, 313, 317, 331, 337, 347, 349, 353, 359, 367, 373, 379, 383, 389, 397, 401, 409, 419, 421, 431, 433, 439, 443, 449, 457, 461, 463, 467, 479, 487, 491, 499.

To help us understand factoring let's discuss divisibility rules.

Divisibility rules

The following rules are helpful for finding the factors of a number (we shall deal only with integers):

A number (integer) is divisible (*i.e.*, divides with a remainder of 0) by:

- **2** if the number is even (ends with 0, 2, 4, 6, or 8);
- **3** if the sum of digits in the number are divisible by 3;
- **4** if the last two digits are divisible by 4;
- **5** if the last digit is 0 or 5;
- **6** if it is even and divisible by 3;
- 7 if 9 times the units digit subtracted from the remaining digits of the number is divisible by 7
- **8** if the last three digits are divisible by 8;
- **9** if the sum of digits is divisible by 9.

Factoring integers completely

The *fundamental theorem of arithmetic* or *unique factorization theorem* states that every positive integer greater than 1 can be written as a product of prime numbers in only one way. For example, $2{,}310 = 2\times3\times5\times7\times11$. There are no other possible factorizations of 1,310 into prime numbers. So, how does one actually

* Refer to the multiplication table: any cell within the table is divisible by its row and column headers.
** The natural numbers are the "counting numbers:" 1, 2, 3, 4, ...

factor a given integer? We'll look at two methods: the *factor tree* and a *division method of factoring*. The diagram on the left shows a factor tree for the integer 1,890. The prime factors are at the tips of the "branches." When the same prime factor appears more than once ("3" in the example) it is customary to show it raised to the appropriate power ("3 ^{3}"). The *division method of factoring*, the same process, is easily done by hand as shown below.

Factor 1,890 completely

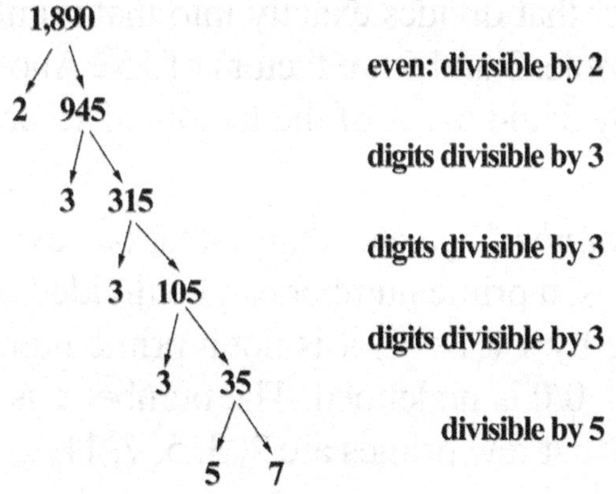

Prime factors: 2, 3, 3, 3, 5, 7. $2 \times 3^3 \times 5 \times 7 = 1{,}890$

Factor 1,890 completely

```
2 |1,890
3 | 945
3 | 315
3 | 105
5 |  35
      7
```

Prime factors: 2, 3, 3, 3, 5, 7.

$2 \times 3^3 \times 5 \times 7 = 1{,}890$ ✓

Coefficient

A *coefficient* is a factor in a term of a polynomial or mathematical expression; it is usually a number, but may be any expression. For example in the quadratic expression $ax^2 + bx + c$, constants a and b are coefficients but c, although a constant, is not a coefficient.*

Composite number

A *composite number* has one or more factors other than the number itself or 1. In other words a composite number is not prime. In fact, except for 1 and 0 which are neither prime nor composite, all whole numbers that are not prime are composite. A composite number is a positive integer that can be formed by multiplying two smaller positive integers. Every positive integer is either composite, prime, or the unit 1. For example, the integer 15 is composite because it is the product of he two smaller positive integers 3 and 5.

The composite numbers** up to 100 are (verify some of these by using the divisibility rules):

4, 6, 8, 9, 10, 12, 14, 15, 16, 18, 20, 21, 22, 24, 25, 26, 27,
28, 30, 32, 33, 34, 35, 36, 38, 39, 40, 42, 44, 45, 46, 48, 49,
50, 51, 52, 54, 55, 56, 57, 58, 60, 62, 63, 64, 65, 66, 68, 69,
70, 72, 74, 75, 76, 77, 78, 80, 81, 82, 84, 85, 886, 87, 88, 90,
91, 92, 93, 94, 95, 96, 98, 99, 100.

If a number is divisible by two or more factors it is also divisible by the product of

* One could argue that c is a factor of "1" and vice-versa because $c = 1 \times c = c \times 1$. Knowledge of this fact is useful but not needed here.

** Check some of these out using the divisibility rules.

these factors. Refer to the factor tree on the previous page; every line shown below 1,890 has two numbers which are factors, and the product of the numbers in any line is the number shown immediately above.

Common factor

Any number that is a factor of two or more numbers is a *common factor* of those numbers. If a set of numbers has two or more common factors than one of them must be the greatest.

Greatest (or Highest) Common Factor (GCF or HCF)

The *greatest common factor* (GCF) of two or more numbers is the largest common factor of those numbers. For example, the factors of 30 are 2, 3, and 5; the factors of 390 are 2, 3, 5, and 13, therefore the common factors are 2, 3, and 5. The largest of these is 5, the GCF of 30 and 390. In grammar school when we reduced a fraction to lowest terms we were simply finding the GCF of the numerator and denominator and then dividing both numerator and denominator by that GCF.

To find either the *least common multiple* (LCM) or greatest common factor (GCF) of two numbers, you always start out the same way: first, find the *prime factorizations* of the two numbers; next, put the factors into a neat grid of rows and columns. Heres how it works:

Example: Find the GCF and LCM or 2,490 and 3,150.
As shown to the left, the factors are

$$2{,}940 = 2\times2\times3\times5\times7\times7 \quad 3{,}150 = 2\times3\times3\times5\times5\times7$$

Now arrange the factors for each number in columns as shown below.

	Factors	Product
2940:	2 2 3 5 7 7	
3150:	2 3 3 5 5 7	
Common factors:	2 3 5 7	$2\times3\times5\times7 = 210$ (GCF)
Factors:	2 2 3 3 5 5 7 7	$2\times2\times3\times3\times5\times5\times7\times7 = 44{,}100$ (LCM)

2 | 2,940	2 | 3,150
2 | 1,470	3 | 1,575
3 | 735	3 | 525
5 | 245	5 | 175
7 | 49	5 | 35
7	7

This orderly listing does most of the work for you! The GCF is the largest number that will divide into both 3490 and 3150. In other words it's the number that is the product of all of the common factors. So the GCF is the product of the factors that are common (same column of the table) to both 2940 and 3150. All of this is displayed in the table above based on finding the factors, here using the division method as shown to the left.

The Least Common Multiple (LCM) is the product of each factor of both numbers; however, factors common to both numbers are counted only once, as shown above in the table. Note that the *Least Common Denominator* (LCD) of a set of fractions is simply the LCM of the denominators.

Relatively prime integers

Two integers ***a*** and ***b*** are said to be *relatively prime*, *mutually prime*, or *co-prime* if the only positive integers (factors) that divides both of them is 1. Thus, 3 and 4 are relatively prime , even though 4 is not a prime number.

Factorials

The *factorial*, ($n!$), of a natural number (positive integer) n, is the product of n and all positive integers less than n. For example, counting problems in probability and statistics: given six objects, the number of *combinations* of pairs of these objects is

$$6!/[(6-2)!\times 2!]=6!/(4!\times 2!)=(6\times5\times4\times3\times2\times1)/(4\times3\times2\times1\times2\times1)=15.$$

Factoring algebraic expressions

Distributive law

The *Distributive Law* (*distributive rule*, or *distributive law with respect to multiplication*) involves both multiplication and addition. This is best illustrated by a few examples:

$$3(2x^3-5x^2+x-2)=3\cdot 2x^3-3\cdot 5x^2+3x-3\cdot 2$$

which simplifies to $6x^3-15x^2+3x-6$.

Here, 3 is the factor of the expression in parentheses. The Distributive Law allows you to "distribute" the factor across each term of the expression contained within the parentheses. The Distributive Law may be expressed symbolically:

$$a(b+c-d)=ab+ac-ad.$$

The factor, a, is distributed and becomes a coefficient of each term within the parentheses (b, c and d in the example). What has all of this to do with factoring? Well, factoring is simply the reverse of the Distributive Law.

Example: Factor the following expression:

$$2x^4-4x^3+8x^2+2x.$$

We must find common factors (if any) of each term in the expression. Note that each term is divisible by 2, and also by x. The common factor, then, is $2x$. Now anything divided by itself (except 0) equals 1, so $2x/2x=1$. So what? Let's multiply the entire expression by $2x/2x$ or equivalently, $2x(1/2x)$, and apply the Distributive Law:

$$2x(1/2x)(2x^4-4x^3+8x^2+2x)$$

Distribute $1/2x$ over each term within the parentheses; that is, divide each term by $2x$ to get

$$2x(2x^4/2x-4x^3/2x+8x^2/2x+2x/2x)$$
$$=2x(x^3-2x^2+4x+1).$$

To check our work we distribute $2x$ over the expression within the parentheses to get the original expression.

This looks complicated because it took a lot of explaining, but in reality one can easily factor many expressions mentally - just look for coefficient and variables that are common to each term* and simply "pull them out." For this example we see that $2x$ is common to each term so we pull it out (which is the same as dividing each term with the parentheses by $2x$) to get the result.

Algebraic patterns

Some algebraic expressions display recognizable patterns that allow one to factor them quickly. Before proceeding further let's see what it means to "factor a polynomial." Factoring an algebraic expression "completely" yields a product of factors, none of which can be factored further. We already have done this for integers.

Example: Consider the rather formidable-looking 5th-degree polynomial

$$f(x) = 2x^5 - 6x^4 - 2x^3 + 6x^2$$

Notice that each term has an even coefficient, therefore is divisible by 2, and that each term is also divisible by x^2. So, let's factor $2x^2$ from the right side of the equation to get

$$f(x) = 2x^2(x^3 - 3x^2 - x + 3)$$

Now, is the expression within the parentheses factorable? Rearrange the terms as show below:

$$f(x) = 2x^2(x^3 - 3x^2 - x + 3) = 2x^2(x^3 - x - 3x^2 + 3)$$
$$= 2x^2[x(x^2 - 1) - 3(x^2 - 1)]$$

Notice that $(x^2 - 1)$ is a factor common to the terms within the brackets ("[ ]"), so we are able to "pull out" that term:

$$f(x) = 2x^2(x^2 - 1)(x - 3)$$

but we're not finished yet! Look at the factor $(x^2 - 1)$ which can be written as $(x^2 - 1^2)$. This is the recognizable "difference of squares" pattern (see below), so it can be factored into

$$(x + 1)(x - 1)$$

and now we have completely factored $f(x)$. The result is

$f(x)$	$= 2x^5 - 6x^4 - 2x^3 + 6x^2$	[Original equation]
	$= 2x^2(x + 1)(x - 1)(x - 3)$	[Completely factored]

The complete factorization of $f(x)$ is the product of 2, x^2, and the linear factors $(x + 1)$, $(x - 1)$, and $(x - 3)$. The factor x^2 is of course factorable into $x \cdot x$ but it is customary to write a factor only once, raising it to the appropriate power.

Difference of squares

$$(a + b)(a - b) = a^2 - b^2$$

The *difference of squares* pattern was used in the above factoring example. It

* It's not necessary that a term actually contain the factor, as we shall see.

works both ways. Notice that $(a+b)$ and $(a-b)$ are linear factors that differ only in signs (one is plus, the other minus). Such pairs are called *conjugate pairs* and are encountered frequently in mathematics.

Perfect square

With natural numbers a *perfect square* is the square of any number multiplied by itself, for example, 25 is a perfect square because $5^2 = 25$. The same is true for algebraic expressions; here we shall limit ourselves to binomials. Thus,

$$(a+b)^2 = a^2 + 2ab + b^2, \text{ a perfect square.}$$

Quadratic trinomial

The standard *quadratic trinomial* is the left side of the equation

$$ax^2 + bx + c = 0$$

where a and b are coefficients and c is a constant.* The quadratic trinomial factors into two linear terms that can be found using several techniques, including factoring by...

inspection (for easy problems),
grouping,
completing the square,
the box, or table, method
and the quadratic formula: $x = \dfrac{-b \pm \sqrt{b^2 - 4ac}}{2a}$

Quadratics are covered in detail in Section 4 (*Quadratics*). There are always two solutions to the quadratic equation. The quadratic formula *always* works and so should be memorized.

Sum of squares and the perfect-square binomial

Let's square the binomial** $(a+b)$

$$(a+b)^2 = a^2 + 2ab + b^2$$

which, when rearranged, gives an important pattern to memorize, the *sum of squares* pattern:

$$a^2 + b^2 = (a+b)^2 - 2ab.$$

Sum and difference of cubes

$$a^3 + b^3 = (a+b)(a^2 - ab + b^2)$$
$$a^3 - b^3 = (a-b)(a^2 + ab + b^2)$$

Note: the operation sign between the cubes is identical to that between terms of the binomial and opposite to that between a^2 and ab of the trinomial (+ and –, or – and +).

* **The term "quadratic" comes from the Latin word** *quadratus* ***which means "squaring."***

** Equivalently, expand the binomial $(a+b)^2$

Expanding - the opposite of factoring

The Binomial Theorem

Note that in the expansion of the perfect square $(a+b)^2 = a^2 + 2ab + b^2$ the coefficients of the trinomial terms are 1, 2, and 1. Now let's expand $(a+b)^3$:

$$(a+b)^3 = (a+b)(a+b)(a+b)$$
$$= a^3 + 3a^2b + 3ab^2 + b^3$$

The coefficients of the expansion terms are 1, 3, 3, 1. Compare these to the coefficients of $(a+b)^2$:

1 2 1
1 3 3 1

There is a definite pattern here which leads to the solution of expanding $(a+b)^n$ using the *binomial theorem* which states that it is possible to expand the polynomial $(a+b)^n$ into a sum involving terms of the form $ma^c\, b^d$, where the exponents c and d are nonnegative integers with $c + d = n$, and the coefficient m, called the *binomial coefficient*, of each term is a specific positive integer depending on n and c. Let's look at the expansion of $(a+b)^4$:

$$(a+b)^4 = a^4 + 4a^3b^1 + 6a^2b^2 + 4a^1b^3 + b^4 \quad (3\text{-}1)$$

Digression: Recall that any expression to the zero power is equal to one*; that is

$$b^0 = 101^0 = (2xy + 13z)^0 = (\text{anything})^0 = 1$$

and anything to the power of "1" is itself:

$$a^1 = a;\ 101^1 = 101;\ (2xy + 13z)^1 = 2xy + 13z;\ (\text{anything})^1 = \text{anything}.$$

With all of this clear in our minds we can rewrite Equation (3-1):

$$(a+b)^4 = a^4 + 4a^3b^1 + 6a^2b^2 + 4a^1b^3 + b^4 \quad (3\text{-}1a)$$
$$(a+b)^4 = a^4b^0 + 4a^3b^1 + 6a^2b^2 + 4a^1b^3 + a^0b^4 \quad (3\text{-}2)$$

Equations (3-1a) and (3-2) are equivalent; all of the Equation (3-2) terms have the form $ma^c\, b^d$ where the non-negative integer exponents (c and d) of a and b, respectively, follow the pattern $c + d = n$ [$n = 4$ in Equation (3-2)]. The coefficient, m, of each term is called the *binomial coefficient* and has the pattern 1, 4, 6, 4, 1. Let's compare this with the previous examples:

1 2 1
1 3 3 1
1 4 6 4 1

Before proceeding we'll look at the expansion of a and b in the trivial cases where $n = 0$ and $n = 1$:

$$(a+b)^0 = 1 \text{ and } (a+b)^1 = a + b \text{ (remember ?)}$$

which give us (on the right side of the equations), respectively, the coefficient patterns of "1"; and "1 1". Place these patterns on the top of the comparisons of the other equations:

* Sometimes, $0^0 = 1$, other times 0^0 is undefined..

1
1 1
1 2 1
1 3 3 1
1 4 6 4 1

We have built a pyramid containing the first five rows of *Pascal's Triangle*!

n	Binomial Coefficients	$(a+b)^n$
0	1	$(a+b)^0$
1	1 1	$(a+b)^1$
2	1 2 1	$(a+b)^2$
3	1 3 3 1	$(a+b)^3$
4	1 4 6 4 1	$(a+b)^4$
5	1 5 10 10 5 1	$(a+b)^5$

Note that the pattern for each line below the first is that the first and last numbers are always "1" and that each number in between is the sum of the two number directly above it. It is easy to see that the next row (for $n=6$) will represent $(a+b)^6$ and will have the binomial coefficients 1, 6, 15, 20, 20, 15, 6, 1. As an exercise, write out the expansion for $(a+b)^7$ and $(a+b)^8$ using Pascal's Triangle. Note the symmetry of Pascal's Triangle. There are many other interesting features of Pascal's Triangle, all of which are beyond the scope of this section.

Roots, zeroes, and *x*- intercepts

What is meant by "*roots of an equation*," "*zeroes of a polynomial*" and "*x-intercepts of a function?*" They are equivalent phrases, representing the points where a function intersects or touches the *x*-axis; that is, where $y=f(x)=0$. See the figure to the right.

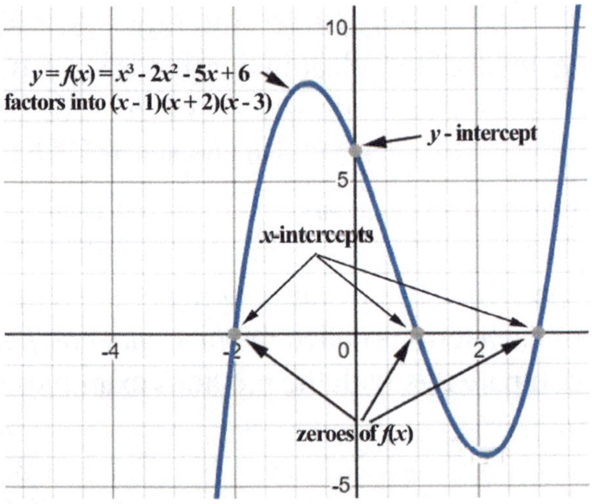

LCM revisited

We have already addressed the LCM of a set of integers as shown below. The same principle applies to polynomials.

	Factors	*Product*
2940:	2 2 3 5 7 7	
3150:	2 3 3 5 5 7	
Common Factors:	2 3 5 7	$2 \times 3 \times 5 \times 7 = 210$ GCM
Factors:	2 2 3 3 5 5 7 7	$2\times2\times3\times3\times5\times5\times7\times7 = 44{,}100$ LCM

GCF and LCM for monomials

As with the finding the GCF and LCM of integers shown above, an orderly listing does most of the work for you!

Example: find the LCM of the monomials $6x$ and $3x^2y$.

	Factors	Product
6x:	2 3 x	
$3x^2y$:	3 $x x y$	
GCF:	3 x	$3 \times x = 3x$
LCM:	2 3 $x x y$	$2 \times 3 \times x \times x \times y = 6x^2y$

The GCF, of course, is the product of the common factors of both monomials and the LCM is the product of all the factors (common factors use only once), as shown in the table on the preceding page. Observe that if you multiply the original monomials (*i.e.*, $6x \times 3x^2y = 18x^3y$) and then divide the product by the GCF (*i.e.*, $18x^3y/3x = 6x^2y$) the result is the LCM. For certain sets of monomials, one way may be easier than the other, but if you employ the table method to find the GCF then the LCM is virtually laid out already.

LCM for polynomials

Example: Find the LCM of the polynomials $x^2 - x - 2$, $x^2 - 4$ and $x^2 - x - 4$.

	Factors			Product
$x^2 - x - 2$:	$(x-2)$	$(x+1)$		
$x^2 - 4$:	$(x-2)$		$(x+2)$	
$x^2 - x - 4$:	$(x-2)$ $(x-2)$			
Common factors	$(x-2)$			$x-2$ GCF
Factors:	$(x-2)$ $(x-2)$ $(x+1)$ $(x+2)$			$x^4+3x^3-2x^2-12x-8$ LCM

It would be unwieldy to find the LCM by multiplying the polynomials together and then divide by the GCF, so the table method shown is probably better.

Example: To find the LCM of $(x-2)$ and $(x+2)$, note that there is no common factor other than 1, so the GCF is the product of the two binomials and the LCM is $(x-2)(x+2)/1 = x^2 - 4$.

Summary of GCF, LCM, & LCD

The greatest common factor (GCF) is the product of all of the common factors of the integers, monomials or polynomials in the set to be considered. The least common multiple (LCM) is the product of all of the factors (common factors considered only once) of the set. Note that when adding (or subtracting) fractions, the least common denominator (LCD) is simply the LCM of all of the denominators. Don't forget: the numerator of each fraction N/D must be multiplied by the LCD/D.

Example: For $a,b,c,d \neq 0$, the expression $a/b + c/d$ has LCD $b \times d$, so the fractions must be changed to $(a \times d)/(b \times d)$ and, $(c \times b)/(b \times d)$.

Rational Root Theorem

Also known as *rational root test*, *rational zero test*, *Rational Zero Theorem*, or *p/q Theorem*, the *Rational Root Theorem* basically states that if a polynomial

$$a_n x^n + a_{n-1} x^{n-1} + \ldots + a_o = 0$$

has rational roots $x = r$, then each such root r is of the form $r = p/q$, p and q relatively prime (e.g. p/q is in lowest terms), then p is an integer factor of a_0 and q is an integer factor of the leading coefficient a_n. If a rational root r is found then the polynomial will be divisible by the linear factor $(x - r)$ using polynomial long division or synthetic division resulting in a polynomial of lower degree whose roots are also the roots of the original polynomial. Note that the theorem does not guarantee the existence of a rational root, but if one exists then the above conditions of factors will hold.

Example: $x^3 + x - 2 = 0$

The only possible rational roots (of integers) are $\pm 1/1 = \pm 1$, and $\pm 2/1 = \pm 2$, and it's easy to confirm that 1 is the root.

The factors of -9 are ±1, ±3, ±9. The factors of 5 are ±1, ±5.

Let's build a table of possibilities:

$p = -9$	$q = 5$	possible roots p/q
±1	±1	1, -1
±1	±5	⅕, -⅕
±3	±1	3, -3
±3	±5	⅗, -⅗
±9	±1	9, -9
±9	±5	9/5, -9/

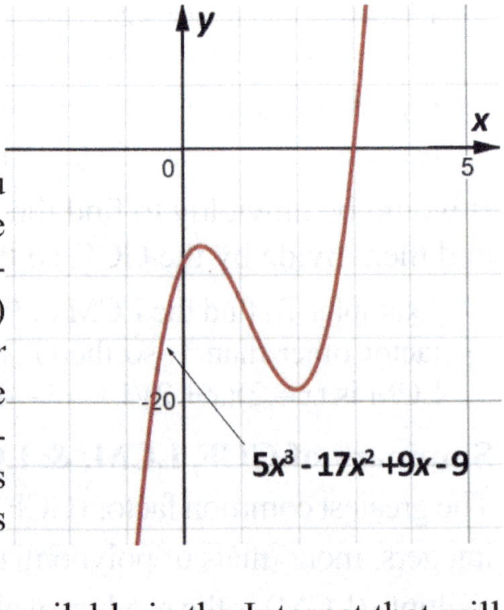

There are twelve possible rational roots. You would have to either "plug in" each candidate to see if the equation holds, or divide the original polynomial by the linear factor $(x - r)$ and look for a remainder of 0, in which case r would be a rational root. See the graph to the right; it turns out that $3/1 = 3$ is the only rational root (easily verified). The other two roots of this equation of degree 3 are not real, as is evident from the graph.

NOTE: There are "rational root" calculators available in the Internet that will do all of the work for you.

•

Polynomial long division

Let's begin with an example:

Example: Divide $3x^3 - 2x + 5$ by $x - 1$.

For this problem $f(x) = 3x^3 - 2x + 5$ and $a = 1$ and we must calculate the quotient $q(x)$ and a remainder $r(x)$, which may be zero. Note first that $f(x)$ has no term of degree 2, *i.e.*, no term containing x^2. The dividend in long division must be in *standard form* with all terms represented, so in this example we insert the x^2 term with a zero coefficient. Always check for variable terms of missing degrees. See Figure 3-1 below. The divisor is $x - 1$ as shown. Long division proceeds as follows:

(1) Divide the first term, $3x^3$ by x to get the first term, $3x^2$, of the quotient.

(2) Multiply each term of the divisor by $3x^2$ and place the product under the dividend with like terms in the same column.

(3) Subtract, term by term, the product from the dividend.

(4) Place the difference resulting from the subtraction under the line as shown. This difference is the new dividend.

(5) Repeat steps (1) through (4) until the difference contains no variable; i.e., is a constant, which may be zero.

The Remainder Theorem shows the relationship between division and multiplication. Just as in arithmetic, when a remainder results from the division of two numbers, the original dividend is determined by multiplying the quotient by the divisor and then adding the remainder. For example, $7/3 = 2$ plus a remainder of 1; multiplying the quotient ("2") by the divisor ("3") gives the product ("6") to which we add the remainder ("1") to get the original dividend ("7").

POLYNOMIAL DIVISION

Divide $3x^3 - 2x + 5$ by $(x - 1)$

Long Division

$$
\begin{array}{r}
3x^2 + 3x + 1 \\
x - 1 \overline{\smash{)}\, 3x^3 + 0x^2 - 2x + 5} \\
\underline{3x^3 - 3x^2 } \\
3x^2 - 2x + 5 \\
\underline{3x^2 - 3x } \\
x + 5 \\
\underline{x - 1} \\
6
\end{array}
$$

Figure 3-1

We have just shown that

$$\textit{dividend} = (\textit{quotient} \times \textit{divisor}) + \textit{remainder}$$

and in terms of the polynomial division

$$f(x) = q(x)\cdot(x - a) + r(x).$$

Hey! Something looks wrong algebraically; why didn't we multiply $r(x)$ by $(x - a)$? Well $r(x)$ is actually the remainder divided by the divisor; in the example, the remainder is $6/(x - a)$ and multiplying that by $(x - a)$ results in $r(x)$. Finally, note that in the example $f(x)$ evaluated at $x = 1$ is equal to 6, the remainder. That is

$$f(x) = 3x^3 - 2x + 5$$
$$f(1) = 3(1)^3 - 2(1) + 5 = 6$$

which is the remainder we found by long division. Clearly, when you divide a polynomial by one of its factors, from the very definition of "factor," the remainder must be zero.

Synthetic Division and the Factor Theorem

Synthetic division is a shortcut that simplifies the process of long division. Here's the example problem again which we shall solve using synthetic division:
Divide $3x^3 - 2x + 5$ by $x - 1$.
Compare $(x - 1)$ to "$(x - a)$;" so, in the example $a = 1$. Synthetic division deals only with the coefficients of the terms of $f(x)$ and, as with long division, to account for missing terms using *place holders* which have a coefficient of zero, as discussed already. Therefore, the coefficients of $f(x)$ are 3, 0, -2, and 5 (the constant 5 is really the coefficient of x^0). Set up the synthetic division as shown in Figure 3-2 below.

Synthetic Division

$$1 \,\big|\; 3 \quad 0 \quad -2 \quad 5$$

Figure 3-2

Another advantage of synthetic division is that there is no dividing or subtracting, only multiplying and adding! The bottom line of a synthetic division gives the coefficients of the quotient and yields the remainder $r(x)$.
Here's the procedure for this example: Refer to Figure 3-3 (next page). The divisor is $(x - 1)$, so position 1 as the divisor. Next, place the coefficients of the decreasing powers of x (3 0 -2 5) under the *vinculum* line, as shown above.

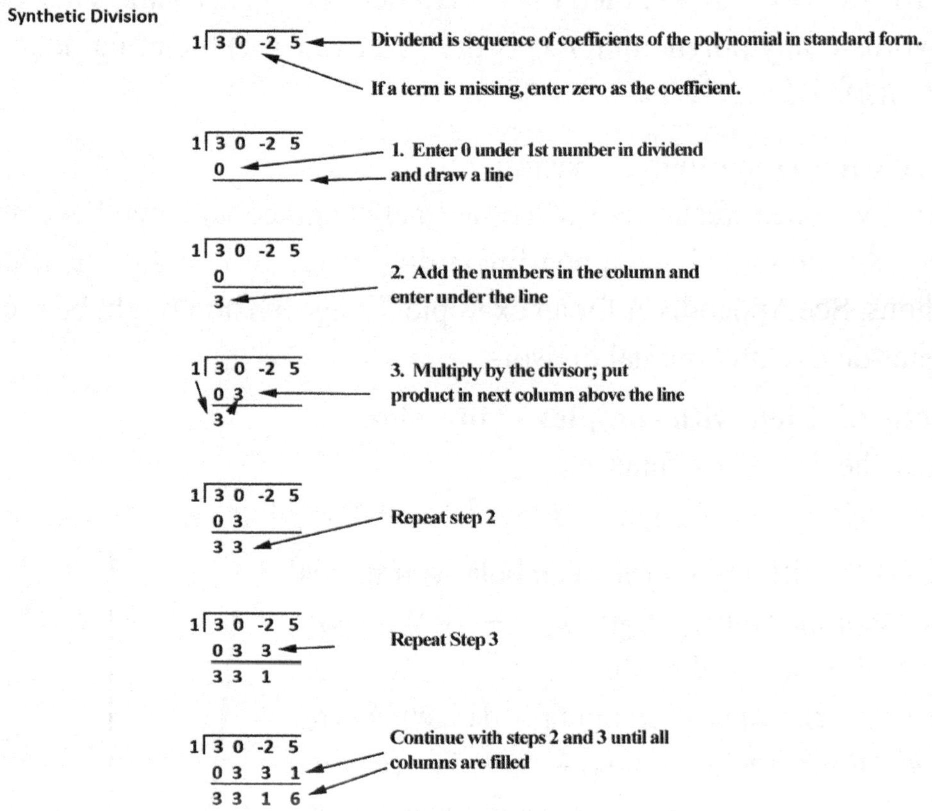

Figure 3-3

1 Insert a zero under the first coefficient ("3");

2 copy the coefficient of the first term and inset it under the line.

3 Multiply 3 by divisor ("1") and place the product under the coefficient of the second term("0").

4 Add the column (0 + 3) and place the sum ("3") under the line. [This is a repeat of Step 2'

5 Multiply the product ("3") of the previous step by the divisor ("1") and place the product under the next coefficient ("2"), then add (-2 + 3 = 1).

6 Multiply the sum ("1") in the previous step by the divisor ("1") and place under the constant ("5") then add
(5 + 1 = 6) and place the sum ("6") in the column under the line. The final result are the coefficients of the quotient:
The first term of the quotient is one degree less than the degree of the first term of the dividend, so the bottom line gives us the quotient: 3 3 1 and the remainder 6.

7. Restoring the variables yields

$$3x^2 + 3x + 1$$

with a remainder of 6, as already found from the more laborious long division.

Factor theorem

If $(x - a)$ is a factor of $f(x)$ then the remainder, $r(x)$, is zero and vice-versa. In other words if any polynomial $f(x)$ is divided by $(x - a)$ the remainder, $r(x)$, is equal to $f(a)$ and vice-versa.

Long division vs synthetic division

Division by a linear factor (e.g. $x - a$) is easily handled with synthetic division; synthetic division involving a non-linear divisor (e.g. $ax^2 + bx + c$) is possible but tedious. See Appendix A for an example. Long-division might be preferable for higher-degree polynomial divisors.

Synthetic division with complex expressions*

Consider the following function:

$$g(x) = x^4 + x^3 + 2x^2 + x + 1$$

the graph of which resembles a parabola with no real zeros. One of the factors of g(x) is $(x - \boldsymbol{i})$. We know, of course, that one of the other factors of g(x) is the conjugate $(x + \boldsymbol{i})$, which when multiplied together is the *difference of squares* $(x^2 - [-1]) = (x^2 + 1)$; we could then divide this into g(x) to get another quadratic, *etc.* However, to illustrate *synthetic division by a complex number* we shall use the root $\boldsymbol{i}$, and then $-\boldsymbol{i}$, as divisors:

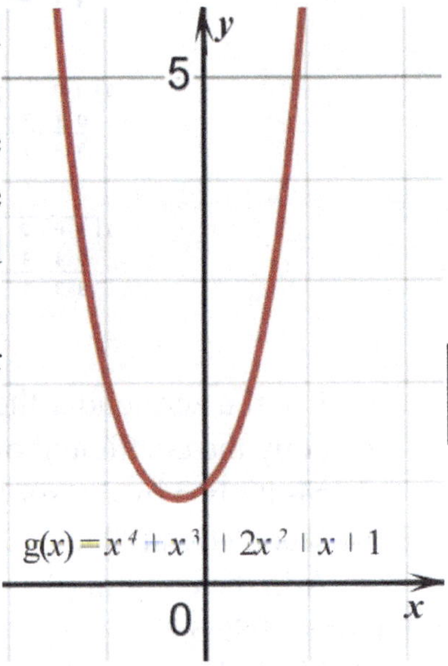

$\boldsymbol{i}$	1	1	2	1	1
		$\boldsymbol{i}$	$-1+\boldsymbol{i}$	$-1+\boldsymbol{i}$	$\boldsymbol{i}^2 = -1$
	1	$1+\boldsymbol{i}$	$1+\boldsymbol{i}$	$\boldsymbol{i}$	0

next, divide the above quotient by the conjugate $-\boldsymbol{i}$

$-\boldsymbol{i}$	1	$1+\boldsymbol{i}$	$1+\boldsymbol{i}$	$\boldsymbol{i}$
		$-\boldsymbol{i}$	$-\boldsymbol{i}$	$-\boldsymbol{i}$
	1	1	1	0

and this quotient is $x^2 + x + 1$ which factors into

$$\left[x - \left(\frac{-1+\sqrt{3}i}{2}\right)\right]\left[x - \left(\frac{-1-\sqrt{3}i}{2}\right)\right]$$

Let's do another example: divide a complex number by its factors using synthetic division.

$$f(x) = x^4 + 3x^3 - 2x^2 - 9x - 3 - 4(x^2 + 3x + 1)\boldsymbol{i}.$$

The real and imaginary parts of f(x) are clear; however, in order to set up this function for synthetic division by one of its zeros, $x - (2 + i)$, we must first collect terms and identify the coefficients of each power of x:

$$f(x) = x^4 + 3x^3 - (2 + 4i)x^2 - (9 - 12i)x - 3 - 4i.$$

Now, set up the synthetic division:

$2 + i$	1	3	$-2 - 4i$	$-9 - 12i$	$-3 - 4i$
		$2 + i$	$9 + 7i$	$11 + 13i$	$3 + 4i$
	1	$5 + i$	$7 + 3i$	$2 + i$	0
	①	②	③	④	⑤ ← [steps]

Steps:

1 Bring down the "1"

2 Multiply $(2 + i)$ by 1 and add to get $(5 + i)$

3 Multiply $(2 + i)$ by $(5 + i)$ to get $(9 + 7i)$ then add to $(-2 - 4i)$ to get $(7 + 3i)$.

4 Multiply $(2 + i)$ by $(7 + 3i)$ to get $(11 + 13i)$ then add to $(-9 - 12i)$ to get $(2 + i)$.

5 Multiply $(2 + i)$ by $(2 + i)$ to get $(3 + 4i)$ which, when added to its opposite, equals zero, and so $(2 + i)$ is indeed a factor of f(x).

That was easy! Now let's divide this quotient

$$x^3 + (5 + i)x^2 + (7 + 3i)x + 2 + i$$

by the conjugate $(-2 - i)$ of the original divisor above, again using synthetic division:

$-2 - i$	1	$5 + i$	$7 + 3i$	$2 + i$
		$-2 - i$	$-6 - 3i$	$-2 - i$
	1	3	1	0
	①	②	③	④ ←[steps]

Steps:

1 Bring down the "1"

2 Multiply $(-2 - i)$ by 1 and add to get 3

3 Multiply $(-2 - i)$ by 3 and add to get 1

4 Multiply $(-2 - i)$ by 1 and add to get a zero remainder.

The quotient is $x^2 + 3x + 1$, a quadratic with real roots.
We see that f(x) has four roots, as expected, and that two of these are real and two imaginary:

$$f(x) = x^4 + 3x^3 - 2x^2 - 9x - 3 - 4(x^2 + 3x + 1)\boldsymbol{i}$$

$$= (x - 2 - \boldsymbol{i})(x + 2 + \boldsymbol{i})$$

•

APPENDIX A

Synthetic division with non-linear divisor

Here's the general process where one polynomial (the dividend) can be divided by a non-linear polynomial (the divisor). The process is the *generalized synthetic division, expanded synthetic division*, or *generalized expanded synthetic division*. If the divisor is a linear term the process is *synthetic division* or more formally, *Ruffini's Rule*. You should already know how to use synthetic division to divide a polynomial by a linear binomial, so we'll refer to that process in the discussion of the generalized synthetic division to divide a polynomial by (in this case) a trinomial. Here's an example:

Example:

Divide $6x^5 + 2x^3 - x^2 + 4x - 2$ by $2x^3 - x - 1$.

Note first that a quintic (degree 5 expression) divided by a cubic will result in a quadratic quotient (the laws of exponents)

• ***Insert place holders**:* identify terms with missing power of x in both divisor and dividend. Insert *zero-coefficient place values* $0x^4$ in the dividend and $0x^2$ in the divisor to yield

$6x^5 + 0x^4 + 2x^3 - x^2 + 4x - 2$ and $2x^3 + 0x^2 - x - 1$.

Before proceeding to apply synthetic division, let's do an old-fashioned polynomial long division, as shown below:

$$
\begin{array}{r|llllll}
 & & & & 3x^2 & +2 & \\
2x^3 - x - 1 & 6x^5 & +0x^4 & +2x^3 & -x^2 & +4x & -2 \\
 & 6x^5 & & -3x^3 & -3x^2 & & \\
\hline
 & 0 & 0 & 5x^3 & +2x^2 & +4x & -2 \\
 & & & 4x^3 & +0x^2 & -2x & -2 \\
\hline
 & & & x^3 & +2x^2 & +6x &
\end{array}
$$

The quotient is $3x^2 + 2$ with the cubic remainder $x^3 + 2x^2 + 6x$ that can be factored into $x(x^2 + 2x + 6)$. As an exercise, verify this by multiplying the quotient by the divisor then adding the unfactored remainder to get the original dividend. So, now we proceed with the problem using generalized synthetic division.

• *First,* insert the sequence of dividend coefficients (here, 6, 0, 2, -1, 4, and -2) as with synthetic division. See Layout 1.

	dividend
	6 0 2 -1 4 -2

Layout 1

There is space for three blank lines below the coefficients, three blank columns in the upper-left quadrant of the diagram, and space for two lines below the horizontal line.

The coefficients of the divisor are 2, 0, -1, and -1.

In regular synthetic division, with a divisor, x + 3 say, the "x" is discarded and the sign of the other binomial term, "3," is reversed so that the synthetic division divisor is "-3" (actually "0 – 3"). Further, if the binomial is not monic, 2x – 5, say, the divisor becomes "0 + 5⁄2"; that is, the other term is divided by the coefficient of the leading term ("x"). Keep this in mind when reading Steps 1 and 2 below.

Steps:

1 The divisor $2x^3 - x - 1$ is not monic so its leading coefficient (2 in this case) is discarded and the signs of the remaining coefficients are reversed. The resulting sequence, (0, 1, 1), becomes the divisor. See Layout 2.

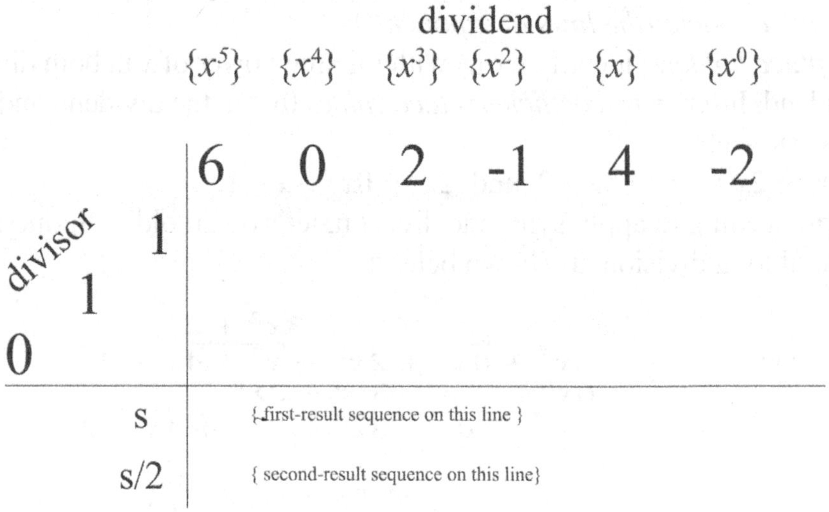

Layout 2

2 Set up the configuration as shown in Layout 2 (the "2" in "s/2" is the *"discarded" leading coefficient of the divisor*).:

Each column of the dividend is labeled with the coefficient of the corresponding variable term: x^5, x^4, *etc*. Of course, the variable associated with the constant term is x^0.

The array includes two lines (labeled "s" and "s/2" in the lower left quadrant and now empty in the lower right quadrant). call them the *first-result sequence* and the *second-result sequence*, as shown.

In the standard synthetic division copy the first number of the divisor sequence to the first position in the answer sequence.

3 Place the number in the dividend's x^5 column ("6") into the first position of the first-result sequence just as in standard synthetic division:

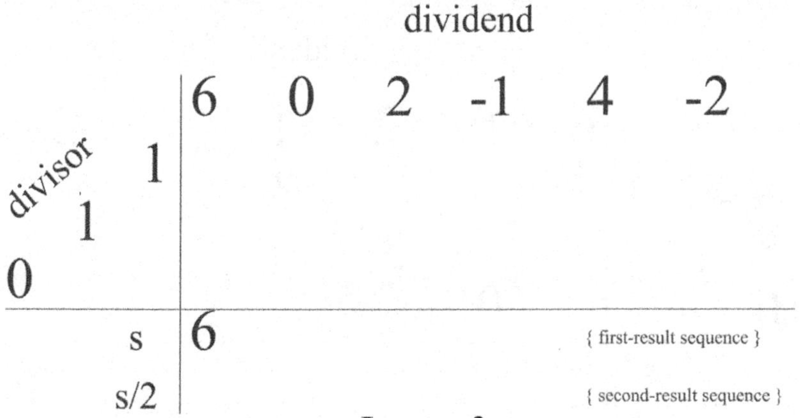

Layout 3

4 Divide the first number of the dividend ("6") by the "discarded" leading coefficient (2, shown in the lower-left quadrant as "s/2."

As with standard synthetic division, divide the coefficient of x into the second term in order to make the divisor a monic binomial. The coefficient of x is "2").

5 Place the result ($6/2 = 3$) into the first position of the second-result sequence (*i.e.*, the second row in the lower-right quadrant of the diagram). This number is the "modified coefficient" of the x^5 term in the original dividend after it has been divided by the leading coefficient of the divisor: See Layout 4.

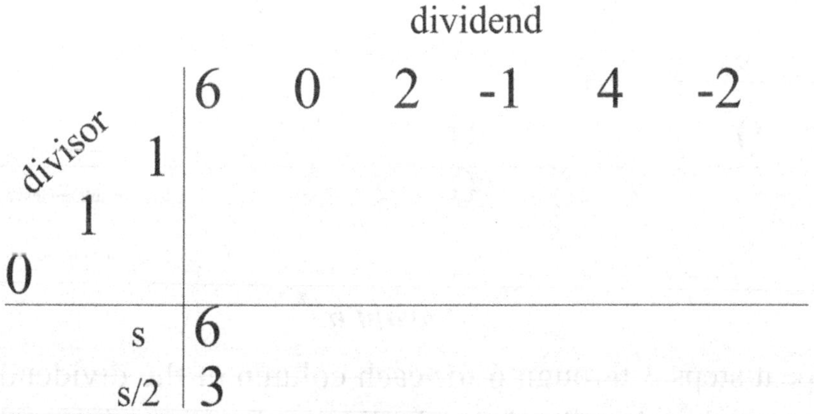

Layout 4

In standard synthetic division we multiply the result by the divisor; do that for each divisor in this example.

Multiply the first entry in the second-result sequence (*i.e.*,"3") by each element of the coefficient sequence of the divisor (0, 1, and 1). Place the products diagonally under the next three dividend columns as shown in Layout 5 below:

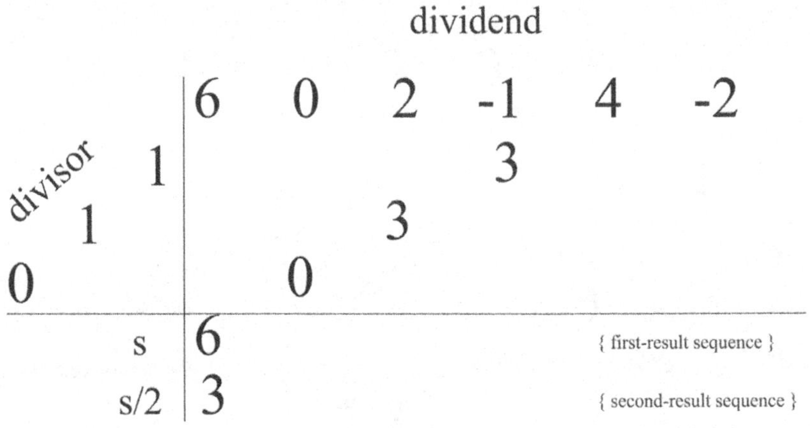

Layout 5

In standard synthetic division, we add the numbers in the column containing the first product.

6 Add the divisor coefficient (0 in this case) to the number in its column of the upper-right quadrant, as shown.

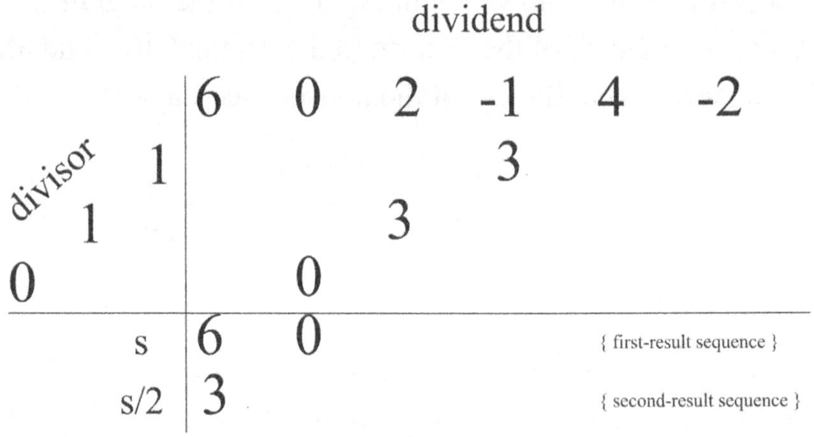

Layout 6

7 Repeat steps 3 through 6 for each column of the dividend. Each step results in a number directly above the line in the upper-right quadrant. Add the numbers in that column, placing the sum in the first-result sequence (under the horizontal line in the lower-right quadrant). Let's finish the example.

The last-dropped number from the first result line (0 in this case) is divided by the leading coefficient of the divisor ("2") to yield a number ($0/2 = 0$) placed on the second result sequence line as shown in Layout 7.

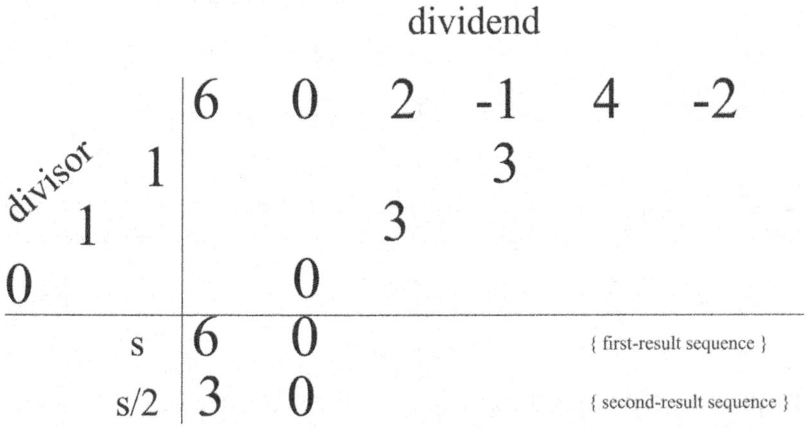

Layout 7

Multiply this result ("0") by the left-hand divisor sequence (0, 1, and 1) to yield products (0, 0, and 0) placed diagonally under the subsequent dividend terms as shown in Layout 8.

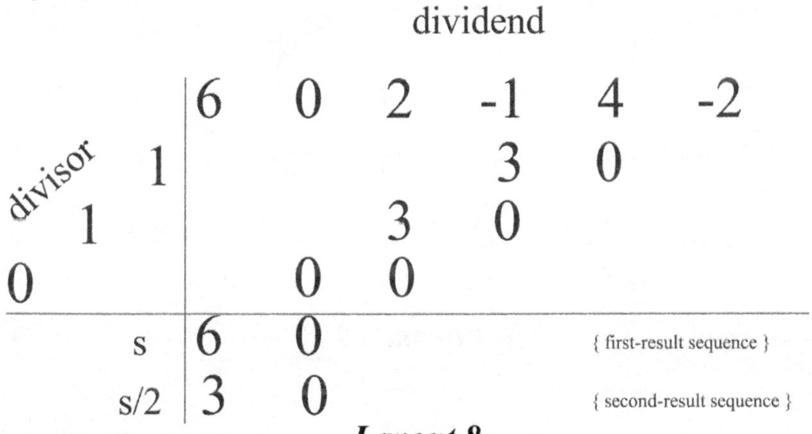

Layout 8

Next comes addition along the subsequent column (the x^3 dividend column, here consisting of 2, 3, and 0), the result of which ($2 + 3 + 0 = 5$) is divided by the leading divisor coefficient ("2") to yield a result of ⁵⁄₂. See Layout 9.

dividend

divisor								
		6	0	2	-1	4	-2	
1					3	0		
1				3	0			
0			0	0				
	s	6	0	5				{ first-result sequence }
	s/2	3	0	5/2				{ second-result sequence }

Layout 9

Finally, the process repeats: multiply ⁵⁄₂ is by the sequence 0, 1, 1 to yield the sequence 0, ⁵⁄₂, ⁵⁄₂ which again is placed diagonally beneath the corresponding dividend terms:

dividend

divisor								
		6	0	2	-1	4	-2	
1					3	0	5/2	
1				3	0	5/2		
0			0	0	0			
	s	6	0	5				{ first-result sequence }
	s/2	3	0	5/2				{ second-result sequence }

Layout 10

Adding the subsequent column (the x^2 dividend column consisting of -1, 3, 0, 0) yields a result ("2"), and because there are three numbers in the first-result sequence (one for each divisor) the remaining columns are added to determine the remainder. This final step is shown in Layout 11.

dividend

divisor						
	6	0	2	-1	4	-2
1				3	0	5/2
1			3	0	5/2	
0		0	0	0		
s	6	0	5	2	13/2	1/2
s/2	3	0	5/2			

quotient (3, 0, 5/2) remainder (2, 13/2, 1/2)

Layout 11

The resulting list of numbers are the left-most three from the second result row and the right-most three from the first result row, namely

$$3,\ 0,\ {}^5\!/_2,\ 2,\ {}^{13}\!/_2,\ {}^1\!/_2\ .$$

In order to determine which of these numbers become the coefficients of the quotient polynomial, first determine how many numbers were in the divisor (upper-left quadrant). There are three numbers (0, 1, and 1), the first three numbers (3, 0, and ${}^5\!/_2$) of the second-result sequence, then, are the coefficients of the quadratic quotient polynomial q(x). Hence, the quotient, q(x), is

$$q(x) = 3x^2 + 0x + {}^5\!/_2 = 3x^2 + {}^5\!/_2$$

The remaining numbers in the first-result sequence of the remaining columns (2, ${}^{13}\!/_2$ and ${}^1\!/_2$) correspond to the coefficients of the remainder polynomial r(x); here,

$$r(x) = 2x^2 + ({}^{13}\!/_2)x + {}^1\!/_2$$

The quotient and remainder combine into the single expression

$$\frac{6x^5 + 2x^3 - x^2 + 4x - 2}{2x^3 - x - 1} = 3x^2 + \frac{5}{2} + \frac{2x^2 + \frac{13}{2}x + \frac{1}{2}}{2x^3 - x - 1}$$

And we are done!

• Let's verify the results: multiply the quotient by the divisor and add the remainder to yield the original dividend polynomial;

$$f(x) = q(x) \times \text{the divisor} + r(x),$$

where q(x) is the quotient and r(x) the remainder:

$$\left(3x^2 + \frac{5}{2}\right)(2x^3 - x - 1)$$

$$= 6x^5 - 3x^3 - 3x^2 + 5x^3 - \frac{5}{2}x - \frac{5}{2} + r(x)$$

$$= 6x^5 + 2x^3 - 3x^2 - \frac{5}{2}x - \frac{5}{2} + 2x^2 + \frac{13}{2}x + \frac{1}{2}$$

$$= 6x^5 + 2x^3 - x^2 + 4x - 2$$

Which was the original polynomial to be divided.

But wait! The long-division quotient was $3x^2 + 2$ and the synthetic-division quotient is $3x^2 + {}^5\!/_2$; how can that be? As an exercise, confirm that both solutions are correct (*Hint*: rewrite $3x^2 + {}^5\!/_2$ as $3x^2 + 2 + {}^1\!/_2$ and add the ${}^1\!/_2$ to the synthetic-division remainder to obtain the long-division remainder).

•

THE MATH TUTOR'S FRIEND
$\frac{-b \pm \sqrt{b^2 - 4ac}}{2a}$
$ax^2 + bx + c = 0$
It's easy !!
$m + n$
$n \times n$
$ax^4 + bx^2 + c = 0$
Mr. Math
QUADRATICS

4. QUADRATIC EQUATION

TABLE OF CONTENTS

•

The Math Tutor's Friend *Book 3: Algebra II*

4. QUADRATIC EQUATION

Standard Form of the Quadratic Equation

The equation $y = f(x) = ax^2 + bx + c$ is the *standard form* of the quadratic (The *monic* form is the standard form with $a = 1$). The graph of a quadratic equation is the *parabola* ("∪" shape). The (real) solutions to a quadratic equation, $f(x) = 0$, are where the parabola crosses or touches the x-axis. More on this later. The sign of a gives the"direction" of the parabola: if a is positive, the vertex is a minimum of the parabola, and if a is negative, the vertex is a maximum (*i.e.*, the parabola is upside down: "∩").

Factoring Quadratics

Factor by inspection

(employing the $m \times n$ and $m + n$ Technique)

This is the easiest method for *monic quadratics*. Simply answer the question "For the quadratic $x^2 + bx + c$, what two integers m and n result in $c = m \times n$ and $b = m + n$?" For example, consider the quadratic $x^2 - x - 12$. Since c is negative (-12), m and n must be of opposite signs, and they must sum to b (-1) so that the absolute value of the negative number prevails by unity. Clearly, 3 and -4 meet the criteria, so the factors are $(x - 4)$ and $(x + 3)$. When factoring quadratics, it is a good policy to spend a short time trying this easy method before becoming involved with more time-consuming factoring techniques. Use a graphing calculator to check your work, or download a graphing app such as *Desmos* to your computer or cell phone.

Complete the square

To square a binomial is simply to multiply it by itself:

$$(x + v)^2 = (x + v) \times (x + v) = x^2 + 2xv + v^2$$

The pattern $x^2 + 2xv + v^2$ is that of a *perfect square.* Let's solve the equation

$$2x^2 + 16x + 32 = 0.$$

First, you should simplify. Note that all the coefficients and the constant are even numbers, so we can simplify the problem by dividing* through by 2 to get

$$x^2 + 8x + 16 = 0$$

which you should immediately recognize as the perfect square $(x + 4)^2$. The "completing-the-square" method of solution changes the quadratic to a perfect square by adding the appropriate constant to both sides of the equation.

* What happens to the "2" ? Look at the general form $n(ax^2 + bx + c) = 0$, where n can be any number. Informally, the value of n determines the "steepness" of the parabola, but for any value of n, the two solutions to the quadratic remain the same. Read about the *Intercept Form* near the end of this book section

Example: Factor by completing the square:

$$3x^2 - 12x + 9 = 0.$$

First note that you can divide through by 3 to get the monic quadratic

$$x^2 - 4x + 3 = 0.$$

(i.e., the coefficient of x^2 is 1).

Now, take half the coefficient of the x term ($-4/2 = -2$) and square it to get 4, then add 4 to both sides of the equation to get

$$x^2 - 4x + 3 + 4 = 0 + 4.$$

Simplify:

$$x^2 - 4x + 4 = 4 - 3 = 1 \text{ [add -3 to both sides]}$$

$$(x-2)^2 = 1.$$

Take the square root of both sides and simplify, to get¶

$$x - 2 = \pm 1$$

from which $x = 3$ or $x = 1$, so the factors are $(x-3)(x-1)$ and

$$x^2 - 4x + 3 = (x-3)(x-1) = 0.$$

Did we change anything when we divided $3x^2 - 12x + 9 = 0$ through by 3? Only the "steepness" of the curve. Read about the *Intercept Form* near the end of this section.

Box method

The box method is a powerful way to factor quadratics, where the leading coefficient is not 1. Here's where the $m \times n$ and $m + n$ technique comes in handy. Let's try it out on

$$12x^2 - 5x - 2 = 0.$$

The first step is to find two integers m and n such that $m \times n = ac = -24$, i.e., $(-2)\times 12$, and $m + n = b = -5$. The product is negative, so that means either m or n (but not both) must be negative. Further, the sum $m + n$ is negative, so the larger absolute value of m or n is negative. The factors (m and n) of 24 are 1×24; 2×12; 3×8; and 4×6. Because m and n must differ by 3, then 3 and 8 are the factors we need and of these, 8 is larger, so $m = -8$ is one factor and $n = 3$ is the other; $m \times n = (-8) \times 3 = -24$ and $m + n = -8 + 3 = -5$, as required, and so the terms we need are $3x$ and $-8x$. Next, we construct a "box" or table to lay out the problem. Place the $12x^2$ term (from the original quadratic) in the upper left-hand box and the constant term, -2, in the lower right-hand box. Now place the terms $3x$ and $-8x$ in the remaining two boxes (it doesn't matter which goes where). The next step it to factor each row and column, as shown below (the sign of each factor is

	$12x^2$	$3x$
	$-8x$	-2

that of the nearest box). The factors are to the left and top: $(3x-2)$ and $(4x+1)$.

	$4x$	1
$3x$	$12x^2$	$3x$
-2	$-8x$	-2

That's it! Problem solved:

$$12x^2 - 5x - 2 = (3x-2)(4x+1) = 0.$$

The quadratic formula

This formula <u>*always*</u> works for the quadratic form $ax^2 + bx + c = 0$. The two solutions are given by the quadratic formula:

$$x = \frac{-b \pm \sqrt{b^2 - 4ac}}{2a}$$

The expression $b^2 - 4ac$ is the ***discriminant*** and provides much information about the underlying parabola. There are three cases:

Case 1: $\boldsymbol{b^2 = 4ac}$; the discriminant is zero and the solutions are real, meaning that the parabola vertex just touches the x-axis at one value of x. There are still two solutions, however; the x-value is said to have "multiplicity of two." The figure to the right shows two parabolas (i.e., two different quadratic equations) the vertices touch the x-axis at $x = -2$ and $x = 1$.

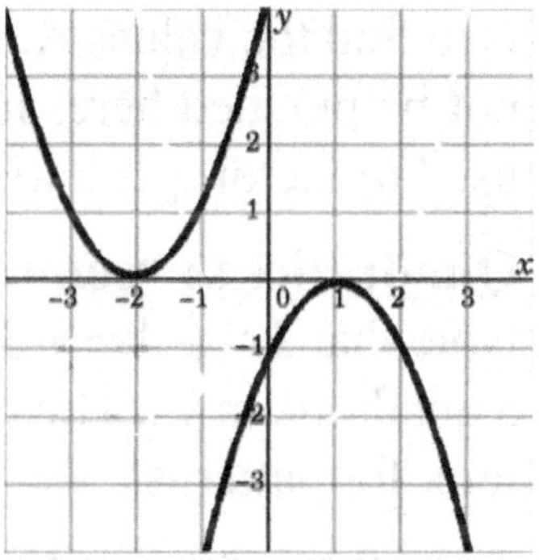

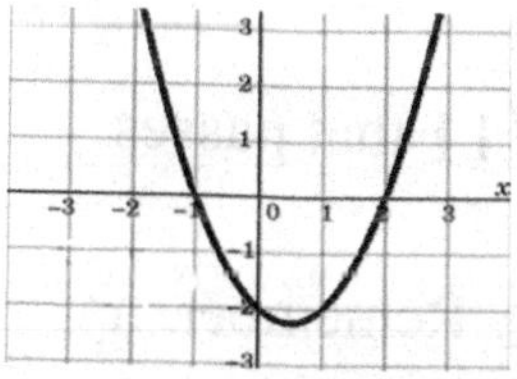

Case 2: $\boldsymbol{b^2 > 4ac}$; the discriminant is positive and the parabola shown to the left crosses the x-axis at two different locations ($x = -1$ and $x = 2$). The solutions are real.

Case 3: $\boldsymbol{b^2 < 4ac}$; the discriminant is negative so the solutions, not real, are of the form $x = u + iw$ and its conjugate $x = u - iw$. The parabola does not touch the x-axis.

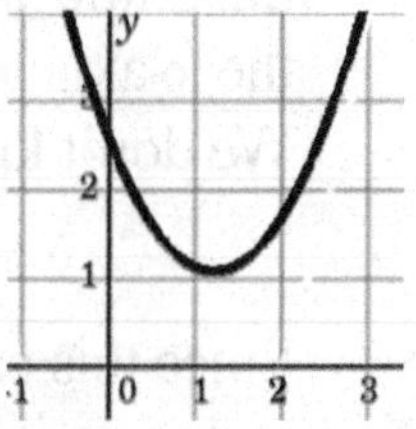

Note: There are two solutions to a quadratic equation and they are either both real or both imaginary.

Factor by grouping

This technique requires you to first group terms with common factors .

<u>Example</u>: Factor $x^2 + 7x + 10$ by grouping. The usual approach is to "split" the middle term into two terms that will permit the factoring. In the example, the middle term is $7x$ which can be split up a number of ways: $x + 6x$; $2x + 5x$; and

$3x + 4x$. Note that among these, the only factors of the quadratic's constant "10" are 2 and 5, so we change the middle, $7x$, to $2x + 5x$ and rewrite the equation as $x^2 + 2x + 5x + 10$. Look at the first two terms. They factor into $x(x + 2)$. The last two terms factor into $5(x + 2)$, so we have a "winner" with a common factor $(x + 2)$:

$$x^2 + 7x + 10 = x(x + 2) + 5(x + 2) = (x + 2)(x + 5).$$

If the coefficient of the "x" term is large you may have to set up a table to get a list of factor pairs. Unless you are asked specifically to factor a quadratic by grouping, it might be easier to use one of the other methods. Remember, the quadratic formula always works.

Factor by "counter-FOIL"

The acronym "FOIL" when multiplying two binomials means "First, Outer, Inner, Last," referring to the terms in the two binomials. Take the quadratic equation:

$$ax^2 + bx + c = 0.$$

This factors into the product of two "linear" binomials:

$$(\mathrm{A}x - \mathrm{B})(\mathrm{C}x - \mathrm{D})$$

which, when multiplied together, yield the original standard form. The problem is to find the values A, B, C, and D, which may be somewhat involved and will not be pursued here. Just be aware that this is one of the methods that can be used to factor quadratics.

Quadratics in "something else"

Note that $3x^4 - 4x^2 + 3$ is a quadratic in x^2 (e.g. let $w = x^2$), and $3x^6 - 4x^3 + 1$ is a quadratic in x^3. The point here is that when analyzing an algebraic expression, look for patterns!

Equation of a quadratic passing through specified points

Example: Find the equation of a quadratic with a zero at $x = \sqrt{11}$ that passes through the point (-1, 20).

Solution: The roots are the conjugates $x = \sqrt{11}$ and $x = -\sqrt{11}$. Remember, a quadratic always has two roots, both real or both imaginary, unless it just touches the x-axis at one point in which case it has one real root with multiplicity two. We don't know yet what the coefficient of x^2 is, so we call it a in our equation:

$$y = a(x - \sqrt{11})(x + \sqrt{11}) = a(x^2 - 11) \text{ [difference of squares]}$$

Since this curve passes through the point (-1, 20), we substitute 20 for y and -1 for x, then write $20 = a[(-1)^2 - 11] = a(-10)$. Solving for a, we get $a = 20/(-10) = -2$, so the solution is

$$y = \mathrm{f}(x) = -2(x^2 - 11) = -2x^2 + 22.$$

Note that there is no "x" term (its coefficient is 0), but we still have a quadratic.*

* This is called a *depressed quadratic* because it contains no linear term. See the discussion in the Appendix B.

Vertex Form of the Quadratic Equation

Recall that for any function $f(x)$, the function $f(x-h)+k$, is simply $f(x)$ translated h units to the right along the x-axis and k units up along the y-axis. Of course, if h and k are negative, the translation is to the left and down, respectively. Now take the parabola $f(x)=x^2$. If we translate the parabola to the right by h units and up by k units, the equation becomes $f(x)=(x-h)^2+k$. Stated another way, the vertex of $f(x)$ was at (0, 0), and is now at (h, k), and this is called the ***vertex form*** of the quadratic equation. So, how do we convert the standard form ax^2+bx+c into the vertex form $a(x-h)^2+k$? An easy way to approach this is to expand the binomial contained in the vertex form and comparing the result with the standard form:

$$y=a(x-h)^2+k=a(x^2-2hx+h^2)+k.$$

Remove the parentheses on the right side: multiply by "a" using the distributive law.

$$y=ax^2-2ahx+ah^2+k.$$

Compare this to the standard form ax^2+bx+c; it is clear that

$$b=-2ah;\text{ and } c=(ah^2+k)$$

from which we can calculate h and k:

$$h=-b/2a \text{ and } k=c-ah^2=c-a(-b/2a)^2=c-b^2/4.$$

The coordinates of the vertex, then, are $[-b/2a, (c-b/4)]$.

Two ways to find the vertex form, $a(x-h)^2+k$.

Method 1: Determine the x-value using the formula $x=-b/(2a)$, then substitute this value for x in the quadratic equation in order to determine the corresponding y-value.

Example: Let $f(x)=x^2+8x-5$

For the quadratic, we have $x=-b/(2a)=-8/(2\times1)=-4$

then compute $y=f(-4)=(-4)^2+8\times(-4)-5=16-32-5=-21$.

So, the vertex of the parabola is at the point (-4, -21); that is, the horizontal translation of x^2 is -4 and its vertical translation is -21, and the vertex form $[f(x)=(x-h)^2+k]$ is

$$y=[x-(-4)]^2+(-21)=(x+4)^2-21.$$

Let $f(x)=(x+4)^2-21$, then $f(-4)=(-4+4)^2-21=-21$, which we already calculated above.

Method 2: Use the "complete the square" method.

Example (same as for Method 1): Put $f(x)$ into vertex form on the quadratic $f(x)=x^2+8x-5$. Take *half* the coefficient of x and square it:

$$(\tfrac{1}{2}\times 8)^2=4^2=16.$$

This will be the third term of the "perfect square" trinomial: $x^2+8x+16$.

Now, add and subtract 16 to $f(x)$: $x^2+8x-5+16-16$. Note that we haven't changed anything; we just added zero (*i.e.*, 16 – 16) to f(x).

Re-arrange $x^2 + 8x - 5 + 16 - 16$ so that the first three terms are the three terms of the perfect square trinomial: $(x^2 + 8x + 16) - 5 - 16$ which becomes $(x^2 + 8x + 16) - 21$. Now $(x^2 + 8x + 16)$ is a perfect square because

$$(x + 4)^2 = x^2 + 8x + 16.$$

Put all of this together to get the vertex form:

$$f(x) = (x - h)^2 + k, \text{ or } f(x) = (x + 4)^2 - 21$$

from which we immediately see that $h = -4$ and $k = -21$, so the vertex is at point (-4, -21) which we also determined in Method 1 above.

Intercept Form of a Quadratic Equation

Is the solution to the quadratic equation unique?

Take the equation $n(x^2 - 3x - 4) = 0$ which factors into

$$n(x - 4)(x + 1) = 0.$$

That is, the zeroes are $x = 4$ and $x = -1$, and the vertex is at $x = 3/2$. But wait! You will get the same results for any integer n (we'll stick with integer coefficients). For example, the following quadratics all have zeroes at $x = -1$ and $x = 4$ with the vertex and *axis of symmetry* at $x = 3/2$:

$$2(x^2 - 3x - 4) = 0$$
$$3(x^2 - 3x - 4) = 0$$
$$10(x^2 - 3x - 4) = 0$$
$$n(x^2 - 3x - 4) = 0 \text{ for all } n.$$

For any integer n, the x-intercepts and x-value of the vertex remain the same; clearly, this is because you can factor out the n-value to arrive at the original factored form (as was done earlier in this section). See the figure below. Note that although the intercepts and x value of the vertices are unchanged, the parabola becomes "steeper" and the y-value of the vertex changes. Notice that n is actually a, the coefficient of x^2 in the quadratic expression.

So, when you factor a monic quadratic, you get only the first, if you will, of

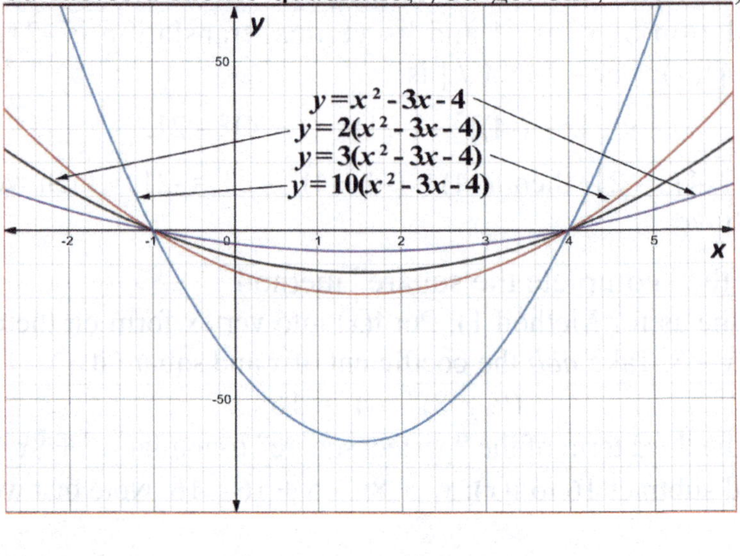

a family of parabolas with identical roots and x-value of their vertices. This is something to be aware of, especially in engineering problems. The above suggests the so-called "intercept form" of the quadratic:

$$a(x-p)(x-q)$$

where a is the coefficient of x^2, as before, and p and q are the x-intercepts (values of x where $y = 0$ and where the parabola crosses the x-axis).

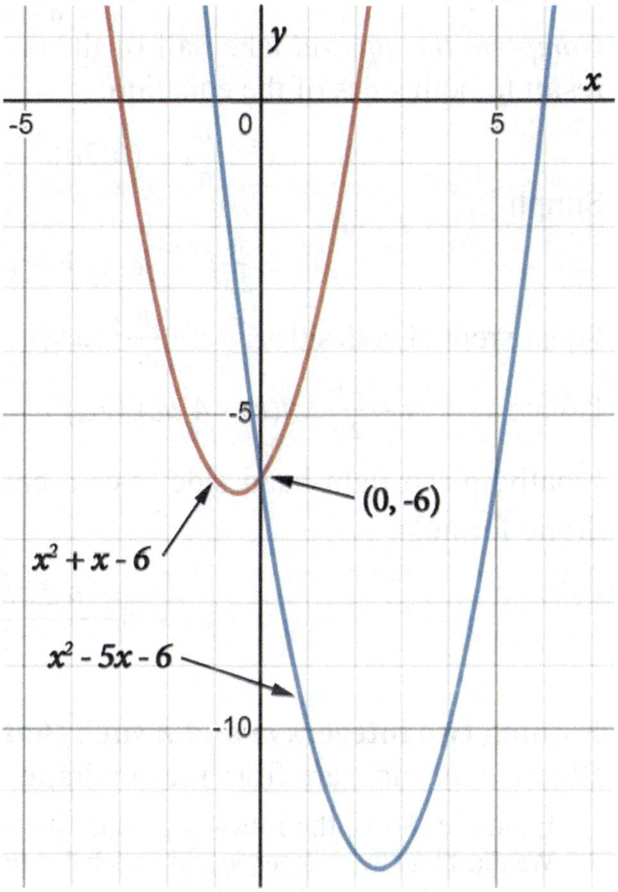

Systems of Quadratic Equations

Solve the following:

$y = (x-2)(x+3) = 0$ (1)

$y = x^2 - 5x - 6 = 0$ (2)

That is, at what points do Equations (1) and (2) intersect?

The first step is to multiply the binomials in order to put them in the trinomial form of the quadratic equation:

$y = (x-2)(x+3) = x^2 + x - 6 = 0$

$y = x^2 - 5x - 6 = 0$

The astute reader will note that the solution is (0, -6); just set x to zero. But let's look at the graphs of these functions (on the right); we are looking for the point(s) of intersection. Sure enough, the intersection is confirmed.

Analysis Summary of the Quadratic Equation

For the quadratic equation:

$$ax^2 + bx + c = 0,$$

If...

$a > 0$ then the graph opens upward;

$a < 0$ the graph opens downward;

$b^2 - 4ac > 0$ then the graph has two x-intercepts;

$b^2 - 4ac = 0$ then the graph has one x-intercept (multiplicity 2);

$b^2 - 4ac < 0$ then the graph has no x-intercept and the solution is imaginary.

The vertex point (axis of symmetry) is at $x = -b/2a$; the y value is found by substituting this in the quadratic formula. The x-intercepts are found by factoring into a product of linear factors or with the quadratic formula. The y-intercept is the value of c in the quadratic equation: $ax^2 + bx + c = 0$.

Derivation of the quadratic formula

Begin with the standard form: $ax^2 + bx + c = 0$.

Add "-c" to both sides: $ax^2 + bx = -c$

Divide by a $x^2 + \frac{b}{a}x = -c/a$.

Complete the square: take half of the coefficient of x and square it. Add the result to both sides of the equation:

$$x^2 + \frac{b}{a}x + (b/2a)^2 = (b/2a)^2 - c/a$$

Simplify:

$$(x + \frac{b}{2a})^2 = -c/a + b^2/4a^2 = (b^2 - 4ac)/(4a^2)$$

Square root of both sides: $x + \frac{b}{2a} = \pm\sqrt{(b^2 - 4ac)}/(2a)$

Isolate x: $x = \frac{-b}{2a} \pm\sqrt{(b^2 - 4ac)}/(2a)$

Finally, place right-hand side over a common denominator to get the quadratic formula.

$$x = \frac{-b \pm \sqrt{b^2 - 4ac}}{2a}$$

Finding two integers *m* and *n* such that *mn* = *ac* and *m* + *n* = *b*

Where a, b, and c are from the quadratic equation.

Example: Solve the following quadratic $5x^2 - 7x - 6 = 0$.
We must find two integer, m and n such that $m \times n = 5 \times (-6) = -30$ and $m + n = -7$. Set up two columns and find the factors of -30 that sum to -7 (either m or n must be negative, but not both – why?):

$m \times n = -30$	$m + n$
-1 × 30	29
1 × -30	-29
-2 × 15	13
2 × -15	-13
-3 × 10	7
→ **3 × -10**	**-7** ←
-5 × 6	1
5 × -6	-1

With experience, you can calculate reasonable numbers in your head, but this example shows a systematic process to help find the answer.

Finding *m* and *n* directly

If the factor method is too difficult, you can compute the values of m and n. We want $mn = ac$, and $m + n = b$, where a, b, and c are the constants in the quadratic equation. Solving the first equation for m, we get $m = ac/n$. Then $m + n$ becomes $(ac/n) + n = b$, from which

$$n^2 - bn + ac = 0,$$

which is a quadratic in n.

Using the quadratic formula we see that the solution is

$$n = (b \pm \sqrt{(b^2 - 4ac)})/2$$

Once we find n, then we can compute m easily.

A note about graphing calculators

A graphing calculator will provide the coordinates of the zeroes of quadratic equations. However, students should know the various analytical methods for dealing with quadratics.

●

APPENDIX B

Depressed quadratic

The standard form of the quadratic equation is

$$f(x) = ax^2 + bx + c = 0 \qquad \text{(B-1)}$$

The solutions, $x_{1,2}$, (or equivalently, x_1 and x_2), to Eqn. (B-1) are given by the familiar quadratic formula

$$x_{1,2} = -\frac{b}{2a} \pm \frac{\sqrt{b^2 - 4ac}}{2a}$$

The two solutions, $x_{1,2}$, to the quadratic are both real (possibly a double root) or both complex conjugates.

Completing the square

The function is a perfect square that can be explained with the following drawing:

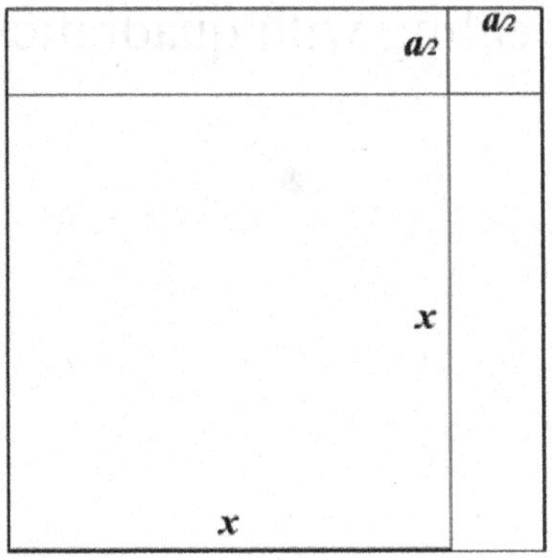

The area of the large square in the figure is the sum of the two squares x^2 and $(a/2)^2$ plus the areas of the two rectangles each of dimension $x \times (a/2)$.

That is,

$$\left(x + \frac{a}{2}\right)^2 = x^2 + ax + \left(\frac{a}{2}\right)^2$$

is a perfect square with the pattern

$$(a + b)^2 = a^2 + 2ab + b^2$$

that should be memorized along with other useful algebraic patterns involving the quadratic such as those in the following table.

QUADRATIC PATTERNS

Perfect square: $(a + b)^2 = a^2 + 2ab + b^2$

Sum of squares: $a^2 + b^2 = (a + b)^2 - 2ab$

Difference of squares: $a^2 - b^2 = (a + b)(a - b)$

Discriminant of the quadratic

The radicand $b^2 - 4ac$ in the quadratic formula is the *discriminant* of the quadratic $ax^2 + bx + c$. The roots of the quadratic equation

$$ax^2 + bx + c = (x - r_1)(x - r_2) = x^2 - (r_1 + r_2)x + r_1 r_2$$

are related to its factors (coefficients of like terms are equal):

$$a = 1;\ b = -(r_1 + r_2);\ \text{and}\ c = r_1 r_2 .$$

The discriminant, Δ, of a quadratic in terms of its roots is

$$\Delta = (r_1 - r_2)^2.$$

For the monic quadratic, $a = 1$; $b = -(r_1 + r_2)$; and $c = r_1 r_2$ and the discriminant* $\Delta = b^2 - 4ac$ expands to

$$\Delta = r_1^2 - 2r_1 r_2 + r_2^2. = (r_1 - r_2)^2$$

Discriminants for the quadratic equation	
Δ	*Roots*
0	one double real root
>0	two distinct real roots
<0	two complex conjugate roots

Depress the quadratic

Begin with the monic quadratic, Eqn. (B-1):

$$f(x) = ax^2 + bx + c = 0 \qquad \text{(B-1)}$$

The leading coefficient, a, is unity (*i.e.*, $a = 1$). Make the substitution $x = t - b/2$ to depress Eqn. (A-1).

$$g(t) = \left(t - \frac{b}{2}\right)^2 + b\left(t - \frac{b}{2}\right) + c = 0 .$$

Expand and simplify to get

$$g(t) = t^2 - \frac{b^2}{4} + c = 0 .$$

g(t) is of the form

$$t^2 + m = 0 \qquad \text{(B-2)}$$

where $m = c - b^2/4$; g(t) is the *depressed quadratic* (for the monic quadratic), so-called because it contains no linear term.

* The discriminant is also expressed as "D."

<u>Example</u>. Consider the quadratic f(x) = $x^2 + 2x + 3$, shown as `"1" in Graph B-1.

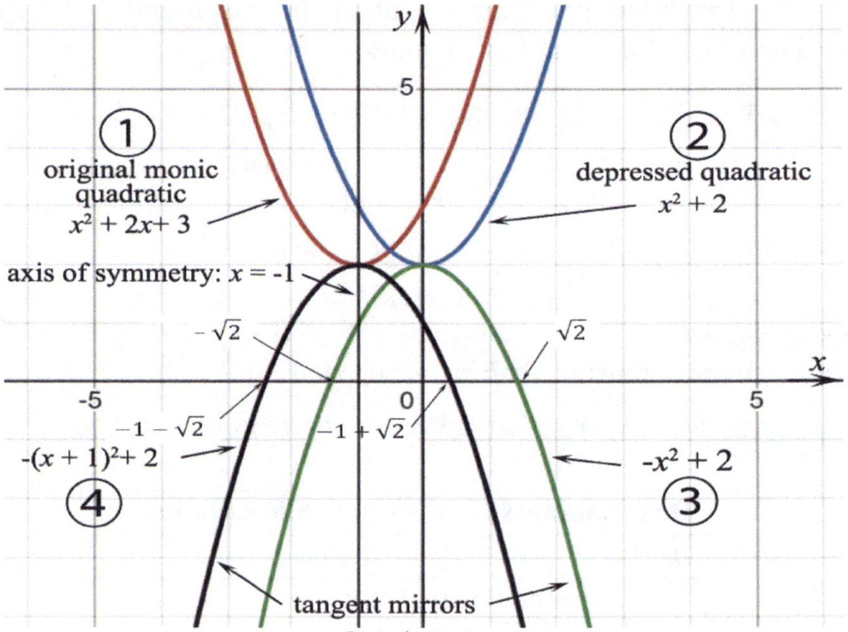

Graph B-1

Here $b = 2$ and $c = 3$, so $m = 2$. The depressed quadratic for f(x) is $f_D(x) = x^2 + 2$. Both forms are plotted in Graph B-1.

Note that Parabolas 1 and 2 in Graph B-1 are identical except for differing axes of symmetry. The axis of symmetry for the original quadratic, $x^2 + 2x + 3$, is at $x = -b/2 = -2/2 = -1$. The linear term for the depressed quadratic is absent, so, $b = 0$ for that function, hence its axis of symmetry is $x = 0$, *i.e.*, the y-axis. In other words, one may plot the depressed quadratic for a parabola simply by making a horizontal translation of the original so that its axis of symmetry coincides with the y-axis; a distance of $b/2$. Parabolas 1 and 2 do not touch the x-axis so we know that the solutions are complex conjugates. Parabolas 3 and 4 illustrate a graphical method to determine the values of the imaginary parts of the solutions. Parabola 3 is $-x^2 + 2$. It is tangent to the depressed quadratic $x^2 + 2$ (Parabola 2). The roots of Parabola 3 intersect the x-axis at x-values $-\sqrt{2}$ and $\sqrt{2}$ which are the real coefficients of $\boldsymbol{i}$ in the complex conjugates $0 - \sqrt{2}\,\boldsymbol{i}$ and $0 + \sqrt{2}\,\boldsymbol{i}$, the solution to the parabola $-x^2 + 2 = 0$. Parabola 4 is a horizontal translation to the left of Parabola 3 by one unit, and is the tangent mirror to the original quadratic, $x^2 + 2x + 3$, with solutions $-1 - \sqrt{2}\,\boldsymbol{i}$ and $-1 + \sqrt{2}\,\boldsymbol{i}$ as shown in Graph B-1. You can verify this by applying the quadratic formula.

Complex conjugate solutions to a depressed quadratic - graphical method

As seen in Graph B-1, the function -f(x) is the opposite (negative) of f(x) and is a reflection of the function f(x) across the x-axis. It is the *mirror image* of f(x). Graph B-2 displays four parabolas.

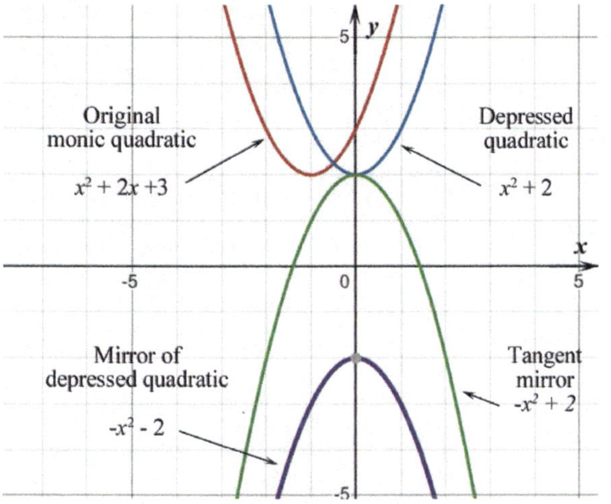

Graph B-2

The original quadratic, f(x), is shown on Graph B-2 along with the depressed quadratic $f_D(x)$, the mirror depressed quadratic $-f_D(x)$ and another inverted parabola. The vertices of the depressed quadratic and its mirror are separated by four units. Vertically translate the mirror function until the two vertices touch; that is, until they are tangent, as shown on the graph. The tangent function, $f_T(x)$ is

$$f_T(x) = -ax^2 - m = -ax^2 - c + \frac{b^2}{4} + 2c = -x^2 + c$$

$a = 1$ and $b = 0$ for the depressed quadratic and $c = 2$ for the examples illustrated in Graphs B-1 and B-2. The remarkable feature about the "tangent inverse" graph $f_T(x)$ is that its axis of symmetry is the real part of the imaginary solutions to f(x) and the x-intercepts are the real coefficients of ***i*** in the complex conjugates of the imaginary solution. For the example shown on the graph, the solutions to the depressed quadratic are

$$x = 0 \pm \sqrt{2}i\,.$$

The roots of the original quadratic

$$f(x) = x^2 + 2x + 3$$

are on its axis of symmetry one unit to the left:

$$x = -1 \pm \sqrt{2}i\,.$$

•

THE
MATH TUTOR'S FRIEND
SYNTHETIC
DIVISION
ASYMPTOTE
It's easy !!
f(x)/g(x)
$\lim_{x\to\infty} \frac{\mathbf{f}(x)}{\mathbf{g}(x)}$
END
BEHAVIOR
Mr.
Math
RATIONAL
FUNCTIONS

The Math Tutor's Friend Book 3: *Algebra II*

5. RATIONAL FUNCTIONS

TABLE OF CONTENTS

•

5. RATIONAL FUNCTIONS

Review of fundamentals

Function notation

The customary method for denoting a function is to use the *function notation* $f(x)$, for example, $f(x) = 3x^4 - 2x^2 + 4x - 2$. To denote another function, one may use $g(x)$, $h(x)$, $P(x)$, $r(x)$, and so on. To evaluate a function $f(x)$ for a given value of x, say 3, the notation is $f(3)$ and x in the function is replaced by the value 3 throughout; thus, for the example,

$$f(3) = 3(3)^4 - 2(3)^2 + 4(3) - 2 = 235.$$

This is sometimes referred to as "plugging a value into the function." The notation $f(x)$ represents a function with the single variable x. If a function has more than one variable, its function notation would be $f(x, y)$ or $f(x, y, z)$, *etc.*

Factor of a polynomial

A *factor* of a polynomial is a constant, another polynomial or a monomial that divides evenly (i.e., no remainder) into the polynomial. As with integers, a polynomial is the product of its factors. For example, the quadratic $2x^2 + 9x - 5$ has factors $(2x - 1)$ and $(x + 5)$. Multiplying the factors yields the original quadratic. Check it out.

Rational functions & dividing polynomials

Definitions:

Algorithm - a sequence of rules to be followed in order to solve a problem.

Asymptote - a line such that the distance between the function and the line approaches zero as both tend to infinity. There are three types of asymptotes: *horizontal asymptote*, *slant* (or *oblique*) *asymptote*, and *vertical asymptote*.

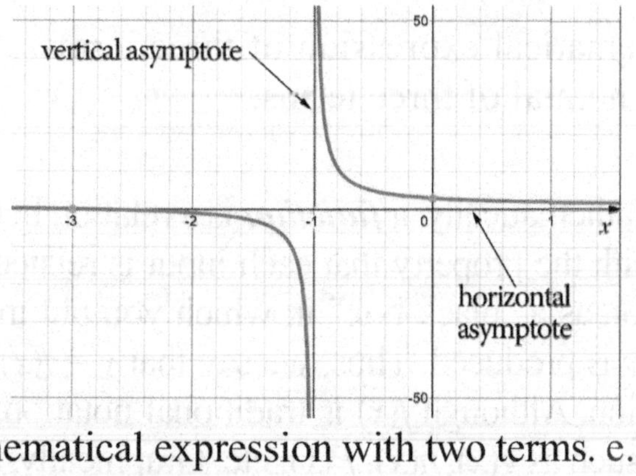

Binomial – a mathematical expression with two terms. e.g. $(x - b)^2$ A binomial is a polynomial of two terms.

Codomain – The set of all possible y values (e.g. real numbers), not necessarily the same as the ***range***.

Domain – The set of all allowable values of x for a function.

End behavior – The value of the rational function that is approached as x becomes extremely large, e.g. approaches infinity.

FOIL (or F.O.I.L.) – A mnemonic acronym for multiplying two binomials; it means "First, Outer, Inner, Last."

Function – a relation between a set of inputs and a set of permissible outputs with the property that each input is related to exactly one output.

Hole – A discontinuity (i.e.,a missing point)* in a function. For example, if b is the x-value for a hole, the function is *undefined* at $x = b$. In a rational function a hole is identified by a factor which is common to both numerator and denominator of the function in factored form, for example, $(x - b)/(x - b)$ signifies a hole at $x = b$.

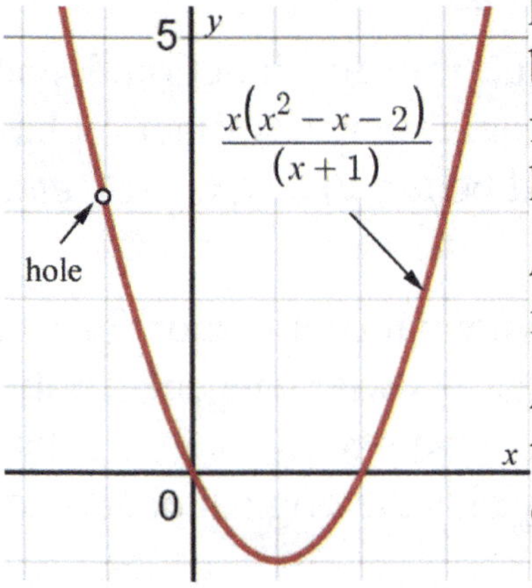

Horizontal asymptote – An asymptote that is a horizontal line, $y = c$

Monomial – A mathematical expression of a single term, e.g. $4x^5yz^3$. Note that a monomial may have consist of many factors.

Polynomial – A mathematical expression of two or more terms, e.g. $4x^5 - 2x^2 + 5$.

Range – The set of y-values for all x in the domain.

Rational number – Any number that can be expressed as a fraction p/q of two integers where q is not 0.

Rational function – two functions expressed as a fraction $P(x)/Q(x)$ where $P(x)$ and $Q(x)$ are functions and $Q(x)$ is not zero.

Trinomial – A mathematical expression of three terms, e.g. $4x^5 - 2x^2 + 5$. A trinomial is a polynomial of three terms.

Functions

What is a function? Mathematically, a ***function*** is a relation between a set of inputs and a set of outputs with the property that each input is related to exactly one output. Think of a function as a "black box" in which you put in a single value of x, and a single value of y is produced. Thus, we say that $y = f(x)$, where the notation $f(x)$ represents a function. Although $f(x)$ is traditional notation, sometime we need to use other notations such as $g(x)$, $h(x)$, $P(x)$, etc. Graphically, a function must pass the "vertical line" test; that is, a vertical line (i.e., parallel to the y-axis) will intersect a function at exactly one point. For example, envision the graph of $y = x^2$. It is "U"-shaped and any vertical line will intersect it at exactly one point. If you tilt

* ***Also called a*** *removable discontinuity.*

that function, however, the vertical line will intersect two points, and the graph is no longer that of a function. A function may be a ***monomial***, ***binomial***, ***trinomial***, or ***polynomial*** (binomials and trinomials are polynomials with special names).

Standard form of a polynomial

The ***standard form*** of a polynomial is simply a conventional way to write with the terms in descending powers of monomials: for example, $2x^3 + x^2 - 3x + 5$ rather than, say, $-3x + 5 + x^2 + 2x^3$, even though both are the same polynomial.

Degree of a polynomial

Let $f(x)$ be a polynomial. The ***degree*** of a polynomial, e.g. $\deg(f(x))$, is the highest degree of its monomial terms. The degree of a term is the sum of the exponents of the variables that appear in it. The term $3^3x^3y^2z$ is of degree 6 (the degree is 6 and not 9 because the rule applies only to variables, i.e., x, y, and z). In this monograph we deal with polynomials of a single variable, usually x, such as $3x^3 - 5x^2 + 2x = 4$, which is a polynomial of degree 3.

Continuous function

A function may be ***continuous*** over an interval [a,b], that is, every value of x in that interval produces a value of y. A function that is not continuous will have "holes" in it or vertical ***asymptotes***. The values of x in these cases are not in the ***domain*** of the function (the function is not defined there). Thus, a function is ***continuous*** when its graph is a single unbroken curve, i.e., it has no jumps or gaps or holes. Intuitively, if a function ***f*** is not continuous you couldn't draw it without lifting your pencil from the paper.

Increasing (Decreasing) function

Moving from left to right on a graph between any two values of x, called an ***interval***, a function will either be increasing (*i.e.*, going "uphill"), decreasing ("downhill") or remain unchanged (or, for a specified interval, may be any combination of the three). One would say that a function increases on an interval of x-values, or decreases on an interval, for example.

Adding, multiplying, and dividing polynomials

Generally, polynomials can be added, multiplied, divided, or raised to a power. Adding is simply an exercise in collecting ***like*** terms (*e.g.*.,having the same power of x). For example, $(3x^3 - 5x^2 + 2x) + (7x^3 - 2x^2 + 4) = 10x^3 - 7x^2 + 2x + 4$. You already have multiplied polynomials (binomials) using the FOIL technique. In general, when you multiply two polynomials, you multiply the first term of the first polynomial with each of the terms in the second polynomial, then the second term of the first polynomial with each of the terms in the second polynomial, and so on. Let's multiply the two polynomials we've just talked about:

$$(3x^3-5x^2+2x)(7x^3-2x^2+4)$$
$$=3x^3(7x^3-2x^2+4)-5x^2(7x^3-2x^2+4)+2x(7x^3-2x^2+4)$$
$$=(3x^3\cdot 7x^3-3x^3\cdot 2x^2+3x^3\cdot 4)-(5x^2\cdot 7x^3-5x^2\cdot 2x^2+5x^2\cdot 4)+(2x\cdot 7x^3-2x\cdot 2x^2+2x\cdot 4)$$
$$=3x^3\cdot 7x^3-3x^3\cdot 2x^2+3x^3\cdot 4-5x^2\cdot 7x^3+5x^2\cdot 2x^2-5x^2\cdot 4+2x\cdot 7x^3-2x\cdot 2x^2+2x\cdot 4$$

which, of course, must be cleaned up using the law of exponents, multiplying integers, collecting *like* terms, and placing in standard form:

$$21x^6-6x^5+12x^3-35x^5+10x^4-20x^2+14x^4-4x^3+8x$$
$$=21x^6-41x^5+24x^4+8x^3-20x^2+8x$$

Division of polynomials is the subject of the next section below.

Rational functions and polynomial division

A ***rational number*** is any number that can be expressed as a fraction *p/q* of two integers, a numerator *p* and a non-zero denominator *q*. Since *q* may be equal to 1; all integers are rational numbers (e.g. 8 = 8⁄1). A real number that is not rational is ***irrational*** (e.g. π). A fraction represents division of the numerator by the denominator – always.

A ***rational function*** *R*(x) consists of two functions expressed as a fraction *P*(x)/*Q*(x), a numerator *P*(x) and a denominator *Q*(x). Any value of *x* which evaluates *Q*(x) to zero is not in the domain of *P*(x)/*Q*(x). Beware of zero denominators! When simplifying a factored rational function, first determine the value(s) of *x* that would cause the denominator to be zero. Such values identify either holes or vertical asymptotes (see the graph on the next page). Make a list of these for later reference. After simplifying, the solution must include that list. For example, consider the following rational function:

$$\frac{x^2+x-6}{x^2-x-2}$$

which factors into

$$\frac{(x+3)(x-2)}{(x-2)(x+1)}$$

Now, before canceling the $(x-2)$ factors, you must determine which values of *x* cause the denominator of the original function to become zero. These values are $x = 2$ and $x = -1$, and so the solution must include the statement "$x \neq -1, x \neq 2$."

$$\frac{x^2+x-6}{x^2-x-2}=\frac{(x+3)}{(x+1)}$$

If you had not analyzed the original function prior to canceling $(x-2)$, you might have missed $x \neq 2$, which identifies a "hole" in the function because the factor $(x-2)$ was also in the numerator of the original function.

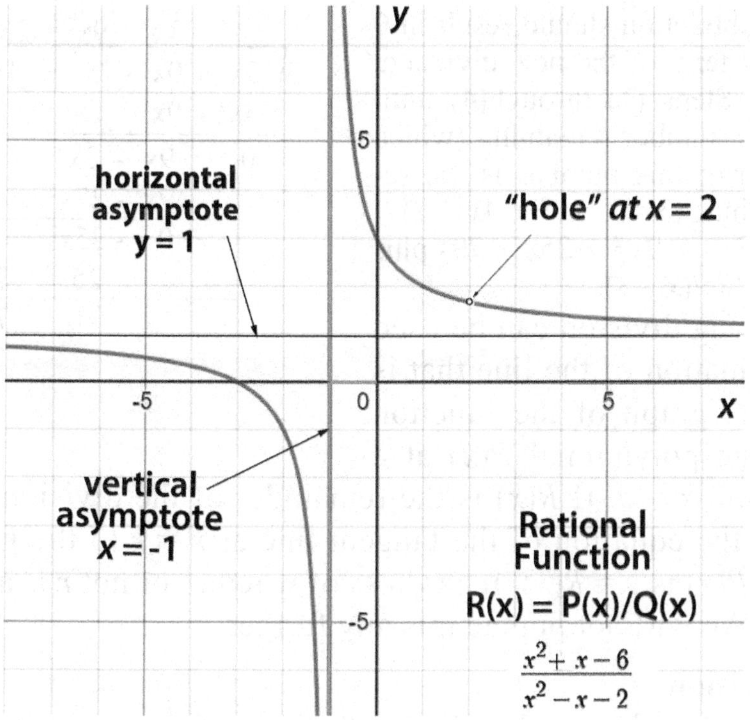

Dividing one polynomial by another

In algebra, ***polynomial long division*** is an algorithm for dividing a polynomial by another polynomial of the same or lower degree, and can be done easily by hand; it is similar to the long division you learned in grammar school. Sometimes, using a shorthand version called ***synthetic division*** is faster, with less writing and fewer calculations. Polynomial long division begins with two polynomials A (the numerator of a rational function, or ***dividend***) and B (the denominator, or ***divisor***, not zero) and yields a ***quotient*** Q and a ***remainder*** R such that

$$A/B = Q + R,$$

and either $R = 0$ or the degree of R is lower than the degree of B. $R = 0$ occurs if and only if B is a factor of A. Thus long division is a means for testing whether one polynomial has another as a factor, and, if it does, for factoring it out. For example, if a root r of A is known, it can be factored out by dividing A by $(x - r)$.

Example: Divide $f(x) = 3x^4 - 2x^2 + 4x - 2$ by the *linear binomial* $(x - 3)$:

Note that $f(x)$, a fourth-degree polynomial, does not contain an x^3 term, thus the term "$0x^3$" must be inserted into the dividend as shown to the left on the next page. Proceed as follows:

(1) Divide the first term, $3x^4$, of the dividend by the first term, x, of the divisor, placing the result, $3x^3$, above in the "cube" column. (2) Then, as in arithmetic long division, you multiply the result, $3x^3$, by each term in the divisor and place the result under the dividend with like terms in the same column.

(3) *Subtract* that line term by term to create a new dividend, as shown.

(4) Each subtraction should result in 0 for the first term of the new dividend.
(5) Repeat Steps (1) through(4) until there is one number remaining (which may be zero); this number is the remainder. Thus, $(3x^4 - 2x^2 + 4x - 2) \div (x - 3) = (3x^3 + 9x^2 + 25x + 79)$ plus remainder $235/(x - 3)$.

$$
\begin{array}{r|l}
 & 3x^3 + 9x^2 + 25x + 79 \\
\hline
x - 3 & 3x^4 + 0x^3 - 2x^2 + 4x - 2 \\
 & 3x^4 - 9x^3 \\
\hline
 & 0 \quad 9x^3 - 2x^2 + 4x - 2 \\
 & \quad 9x^3 - 27x^2 \\
\hline
 & \quad 0 \quad -25x^2 + 4x - 2 \\
 & \quad\quad 25x^2 - 75x \\
\hline
 & \quad\quad 0 \quad 79x - 2 \\
 & \quad\quad\quad 79x - 237 \\
\hline
 & \quad\quad\quad 0 \quad 235
\end{array}
$$

Polynomial long division can be used to find the equation of the line that is tangent to the graph of the function defined by the polynomial $P(x)$ at a particular point $x = r$. If $R(x)$ is the remainder of the division of $P(x)$ by $(x - r)$, then the equation of the tangent line at $x = r$ to the graph of the function $y = P(x)$ is $y = R(x)$, regardless of whether or not r is a root of the polynomial. We will not pursue this any further.

Synthetic division

In order to divide polynomials using synthetic division, you must divide by a *monic linear expression* (*i.e.*, the leading coefficient must be "1"). For example, you can use synthetic division to divide by $x + 4$ or $x - 1$, but it is overly involved to employ synthetic division to divide by, say, $x^2 + 4$ (see Appendix A). If the leading coefficient of the divisor binomial is not a 1, then you must divide by that coefficient to turn the leading coefficient into a 1. For example, $(4x - 1)$ would become $(x - ¼)$ and $(4x - 3)$ would become $(x - ¾)$. If synthetic division doesn't work, then you must use long division.

Synthetic Division of a Polynomial P(x) by a binomial B(x), i.e., $P(x)/B(x)$: First, set the denominator, $B(x)$, equal to zero to find the number to use as divisor. Next, make sure the numerator is written in standard form (descending order of exponents) and if any terms are missing you must fill in the missing term wit a "0." Finally, list only the coefficient in the division problem. For example, the dividend $x^5 - 3x^2 + 2$ must be expanded to $x^5 + 0x^4 + 0x^3 - 3x^2 + 0x + 2$; the dividend of the synthetic division is then written as 1 0 0 -3 0 2.

Example: Divide $f(x) = 3x^4 - 2x^2 + 4x - 2$ by the linear binomial $(x - 3)$ using synthetic division:
First, recognize that $(x - 3)$ is a binomial which may or may not be a factor of $f(x)$. In synthetic division, for the binomial $(x - a)$, the divisor is a; for the binomial $(x + a)$ the divisor is $-a$. So, with $(x - 3)$, the divisor is 3.
The dividend for synthetic division is a list of the coefficients of the variable for the function in standard form, one coefficient for each power of that variable. If a power is missing, then a zero is inserted. This is easier than it sounds; let's look at $f(x)$ above: it is in standard form; that is the terms are in descending

powers of x. Note that there is no x^3 term, so the dividend is 3 0 -2 4 -2.

Steps (see the figure to the right):

(1) Insert the divisor, 3, to the left of the vertical line;

(2) List the variable coefficients and constant to the right of the vertical line;

(3) Bring down the first coefficient to the bottom;

3	3	0	-2	4	-2
		9	27	75	237
	3	9	25	79	235

(4) Multiply the first coefficient by the divisor and place the product in the middle row under the second coefficient;

(5) Add the product to the second coefficient and place in bottom row;

(6) Multiply the result by the divisor and place in the middle row;

(7) Repeat steps (4), (5), and (6) for each remaining coefficient;

(8) The quotient and remainder (which may be zero) appears in the bottom row. The degree of the quotient polynomial is one less than that of the dividend. The last number, which may be zero, is the remainder. If the remainder is zero, then $(x - a)$ is a *factor* of the original polynomial. In the example, the quotient is $3x^3 + 9x^2 + 25x + 79$ with remainder 235. As a bonus, note that the remainder is the value of the function evaluated at the divisor; thus in the problem above, $f(3) = 235$, which is the remainder. This leads us to the Remainder Theorem:

Remainder Theorem

Divide f(x) by $(x - a)$. If the remainder, r, is zero then $(x - a)$ is a factor of f(x); otherwise, f(a) = r. The *Remainder Theorem* addresses the division of a polynomial $f(x)$ by a linear factor $(x - a)$ where a is a constant. If you divide $f(x)$ by $(x - a)$ using *long division*, say, you will get a quotient $q(x)$ and a remainder $r(x)$ which may be zero in which case $(x - a)$ is a factor (divides evenly) into $f(x)$ just as 3 divides evenly into 24. Algebraically, $f(x)/(x - a) = q(x) + r(x)$. If the remainder, r, is zero, then $(x - a)$ is a factor of $f(x)$; otherwise, $f(a) = r$.

The Remainder Theorem shows relationship between division and multiplication. Just as in arithmetic, when a remainder results from the division of two numbers, the original dividend is determined by multiplying the quotient by the divisor and then adding the remainder. For example, 7/3 = 2 plus a remainder of 1; multiplying the quotient ("2") by the divisor ("3") gives the product ("6") to which we add the remainder ("1") to get the original dividend ("7"). We have just shown that

$$\textit{dividend} = (\textit{quotient} \times \textit{divisor}) + \textit{remainder}$$

and in terms of the polynomial division

$$f(x) = q(x)\cdot(x - a) + r(x).$$

Factor theorem

If $(x - a)$ is a factor of $f(x)$ then the remainder, $r(x)$, is zero and vice-versa. In other words if any polynomial $f(x)$ is divided by $(x - a)$ the remainder, $r(x)$, is equal to $f(a)$ and vice-versa.

Behavior of Rational Functions

Bounded polynomial

In mathematics, a function f is said to be *bounded* if the range of its values is bounded. That is, there exists a real number M such that $abs(f(x)) \leq M$ for all x in the domain. This means that if the range of a function does not go beyond a certain value, then the function is bounded. otherwise, the function said to be *unbounded.*

Examples: $(x + 2)^2 + 4$ is bounded (by the value of y at the vertex), and x^3 is unbounded.

Additionally, if $f(x) \leq A$ for all x in X, then the function is said to be *bounded above* by A. On the other hand, if $f(x) \geq B$ for all x in X, then the function is said to be *bounded below* by B.

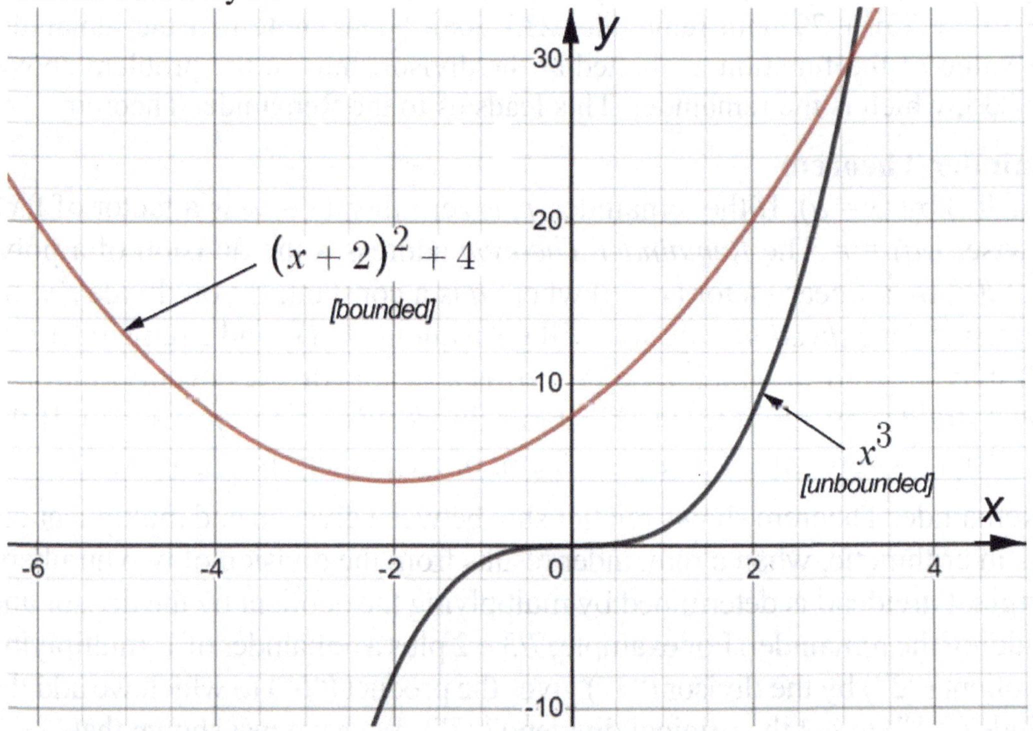

Asymptotes and "Holes"

Already briefly mentioned, an ***asymptote*** is a line such that the distance between the curve and the line approaches zero as they tend to infinity. That is, the larger x becomes, the closer to the asymptote the curve becomes. A ***hole***, (also called *discontinuity* or *removable singularity*) on the other hand, is a value of x where $f(x)$ is undefined. For example:

$$\frac{x^4 + 2x^3 - x^2 - 2x}{x^3 + 2x^2 - 3x}$$

Factoring x from the expressions in both numerator and denominator:

$$\frac{x(x^3 + x^2 - x - 2)}{x(x^2 + 2x - 3)}$$

There is a common factor, x, in both numerator and denominator. Before canceling we note that x/x represents a "hole" in the graph at $x = 0$ because 0/0 is undefined. After canceling x/x, we have

$$\frac{x^3 + 2x^2 - x - 2}{x^2 + 2x - 3}$$

Which, after factoring numerator and denominator, yields

$$\frac{(x^2 - 1)(x + 2)}{(x - 1)(x + 3)}$$

After factoring the "difference of squares" in the numerator we have

$$\frac{(x + 1)(x - 1)(x + 2)}{(x - 1)(x + 3)}$$

The term $(x - 1)$ appears in both numerator and denominator, so there is another "hole" at $x = 1$. Now we have

$$\frac{(x + 1)(x + 2)}{(x + 3)}$$

Setting the denominator equal to zero we see that there is a vertical asymptote at $x = -3$. So, what's the difference between a "hole" and a vertical asymptote? Look at the ratios. If the factor in the numerator is identical to that of the denominator, e.g. x/x, $(x - 1)/(x - 1)$, then there is a hole, because at $x = 0$, we have 0/0 which is undefined and at $x = 1$, $(x - 1)/(x - 1)$ also undefined. For vertical asymptotes, when the denominator becomes zero the numerator does not; this is the case as $x = -3$.

Limits

Function $f(p)$ has limit L: this means $f(x)$ gets closer and closer to L as x moves closer and closer to p. The notation $\lim_{x\to\infty}$ indicates that the value x approaches infinity.

End behavior

Consider the behavior of a rational function as x approaches ± infinity.

When examining a function at large absolute values of x, we look at the limit of the ratio $f(x)/g(x)$ as x approaches positive infinity. That is, $\lim_{x\to\infty} \frac{f(x)}{g(x)}$, or the ratio $f(x)/g(x)$ as x approaches negative infinity. There are three cases to consider that depend upon the relative degrees of the numerator and denominator:

(1) the degree of the numerator is less than the degree of the denominator;

(2) the degrees of the numerator and denominator are equal; and

(3) the degree of the numerator is greater by one than that of the denominator.

Let's look at an example for each of these three cases:

Case 1: *deg*[*f*(*x*)] < *deg*[*g*(*x*)]

$$\frac{x^3 + 3x - 4}{x^4 - 5x + 1}$$

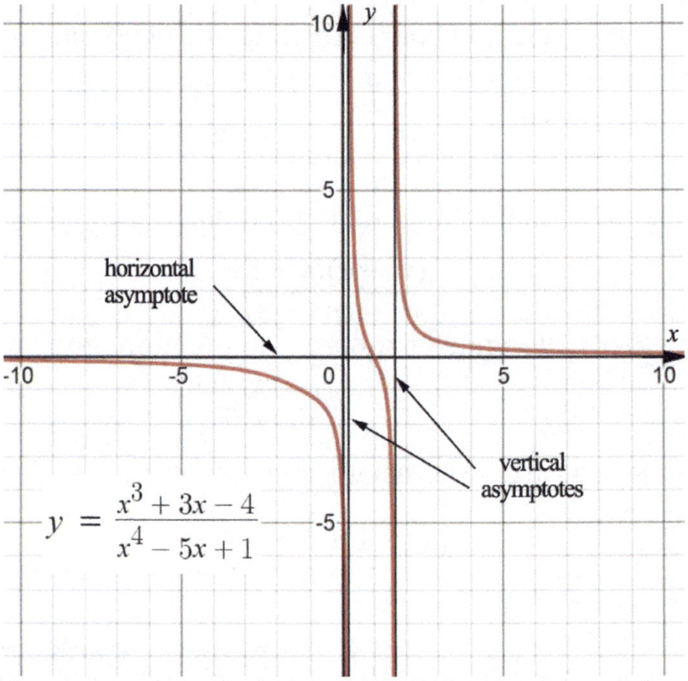

Envision *x* becoming extremely large; if large enough, the last two terms of each function become insignificant so that the function essentially becomes close to

$$\frac{x^3}{x^4}$$

or 1/*x*, which tends to zero as *x* approaches infinity. Now, *x* will never reach zero, so the line $y = 0$ is the *horizontal asymptote* of the original function. This is true even if the coefficients of the leading terms are not equal to 1.

Case 2: *deg*[*f*(*x*)] = *deg*[*g*(*x*)]

$$\frac{6x^3 - 2x + 11}{2x^3 + 15x - 2}$$

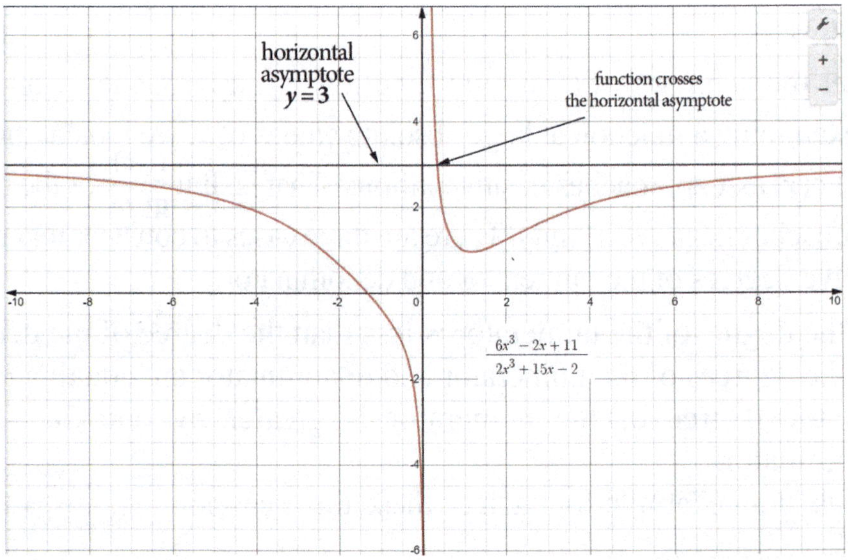

As x approaches infinity, the function will look like $6x^3/2x^3$, which is $6/2 = 3$. So this function has a *horizontal asymptote* which is the ratio of the two leading coefficients. Note from the graph that a function may cross (once) its horizontal asymptote. To determine where this might occur, solve the equation $f(x)/g(x) = k$, where k is the y-value of the horizontal or slant asymptote.

Case 3: $deg[f(x)] = deg[g(x)] + 1$

$$\frac{3x^4 - 2x^2 + 4x - 2}{3x^3 + 9x^2 + 25x + 75}$$

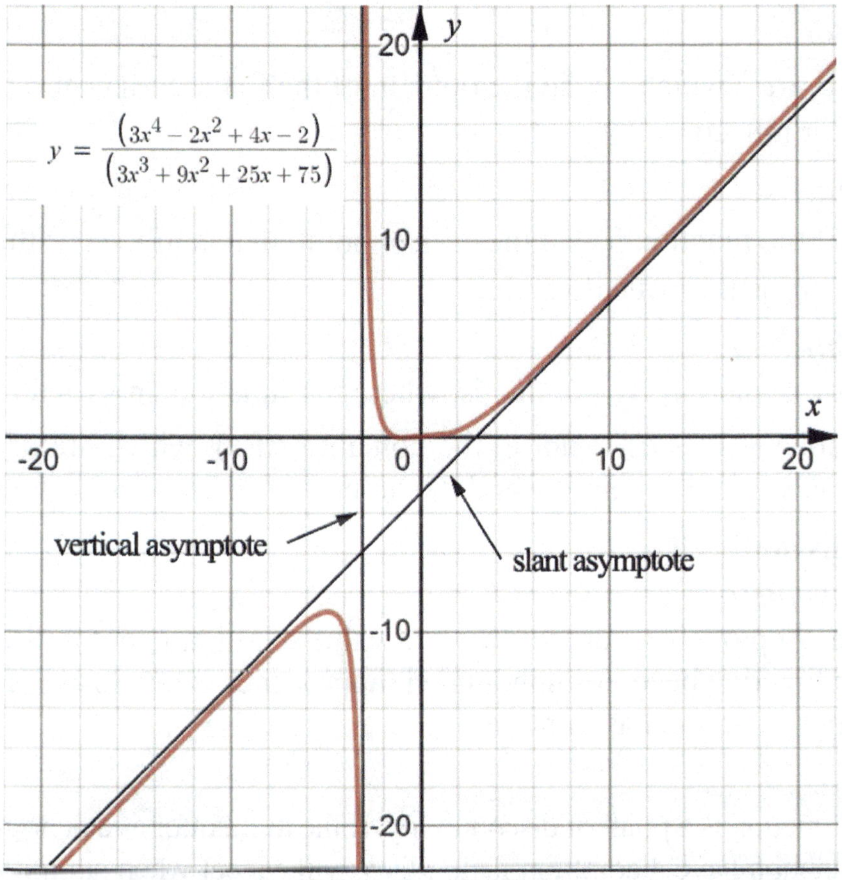

Using polynomial division or synthetic division, we see that the quotient is $x - 3$ plus a remainder. The quotient is a linear equation which gives us a slant (or oblique) asymptote along the line $y = x - 3$, which has slope of one and y-intercept at $y = -3$. There is no vertical asymptote.

Critical points, critical values, zeroes, points of discontinuity

With respect to functions, there seems to be inconsistency in the definitions of the terms *critical point*, *critical value*, and *zero*. Make sure that you understand the definition used in the context of your work. We'll use the following definitions.

* Critical points that are extrema are found using calculus by setting the first derivative of a differentiable function to zero and inflection points are those values of x where the second derivative is equal to zero. Critical points also include values of x in the domain of the function where the derivative does not exist as with absolute value functions, functions with cusps and inflection points of cube roots, for example.

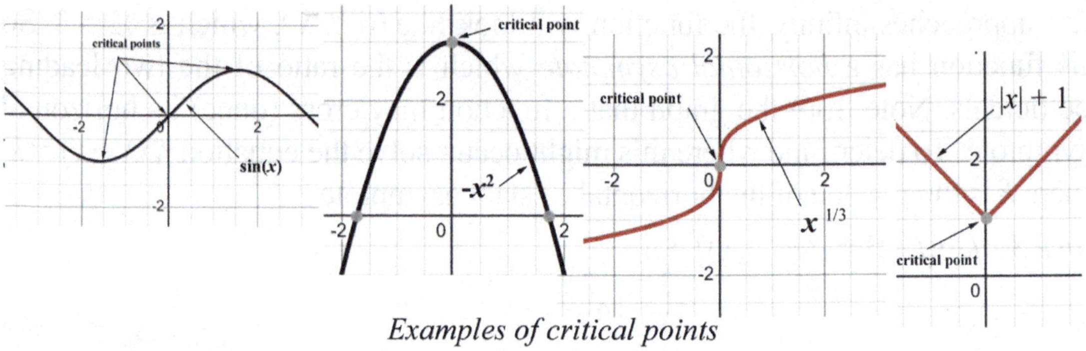

Examples of critical points

critical point

The point (c, f(c)) which lies in the domain of the function f(x) and is a point of *extrema* or inflection.*

critical value

The value of the function f(x) at the critical point; that is, if $x = c$ is the critical point, then f(c) is the critical value.

zero of a function

The value of x where f(x) = 0; that is, where the function crosses the x-axis.

There are other values of x, not in the domain of f(x), which are of interest. These are the vertical asymptote and the hole.

Points of discontinuity

Vertical asymptote

A vertical asymptote is a vertical line at a value of x that is not in the domain of f(x). The function becomes unbounded (approaches ∞ or -∞) as x nears (but never touches) a vertical asymptote.

Hole

A hole of a function is a point of discontinuity. If the numerator and denominator of a rational function have a common factor, they will cancel when simplifying. The value of x that sets the factor to zero defines a hole in the graph at (x, f(x)) and is not in the domain of f(x).

It should be clear from these definitions that vertical asymptotes and holes are not critical values because their locations are not in the domain of the function.

Finding zeroes of a rational function

A rational function is one polynomial divided by another in the denominator. It may be expressed as $f(x) = P(x)/Q(x)$ where $Q(x)$ is non-zero. Just as an integer is a rational number having a 1 in the denominator, so too can $f(x) = P(x)/1$ be a rational function (in such a case, $Q(x) = 1$). Solving rational equations involves several steps:

1) If one side is not zero, then rewrite the equation so that one side becomes zero.

2) Find all of the linear factors of $P(x)$ and $Q(x)$.

3) Find the critical values of each linear factor by equating each to zero and solving for the variable (e.g. x). It will be helpful to mark each value on a number scale.

4) Determine the sign of the original function in each interval formed by the critical values.

5) The solution will be those intervals within which the variable allows $f(x)$ to satisfy the equation.

Example: $f(x) = x^3 - x^2 = 6x$ [Here, $P(x) = f(x)$ and $Q(x) = 1$]
First, re-arrange the equality to make the right side zero by adding -6x to each side:

$$f(x) = x^3 - x^2 - 6x = 0$$

The numerator, $f(x)$, can be factored as follows:

$$x(x^2 - x - 6) \quad \text{[factor out } x\text{]}$$

$$= x(x + 2)(x - 3) \quad \text{[factor the quadratic].}$$

Now by setting each factor to 0, calculate the three ***zeroes***. i.e. values of x that will make $f(x) = 0$. The zeroes are $x = 0$; $x = -2$; and $x = 3$.

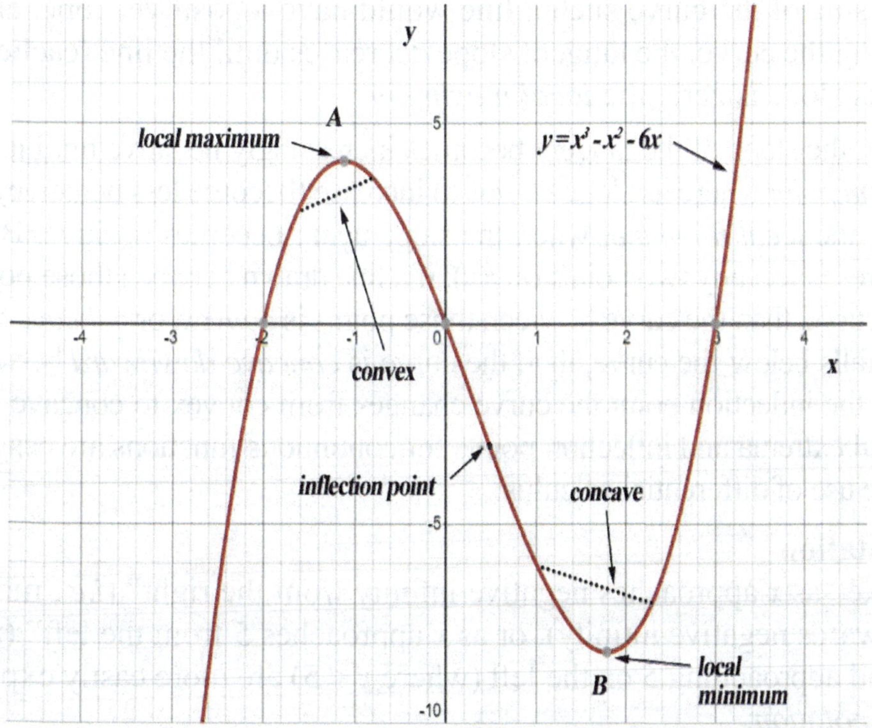

Construct a number line showing these points:

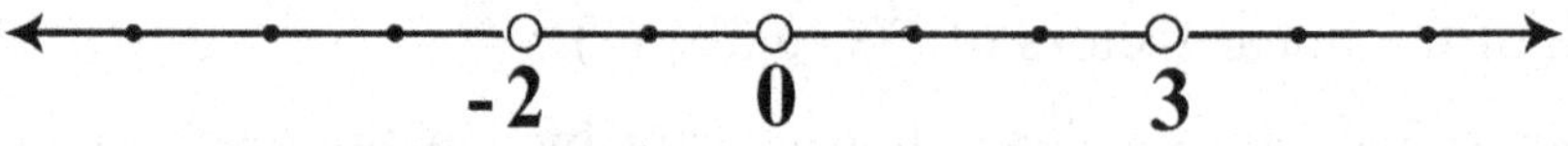

The number line displays four intervals: $(-\infty, -2)$; $(-2, 0)$; $(0, 3)$; and $(3, \infty)$. Next, select a value in each interval and determine the sign of f(x). Select "easy" values such as $x = -3, x = -1, x = 1, x = 4$, to simplify computation. See the table below.

Interval	x	$f(x) = x^3 - x^2 - 6x =$
$(-\infty, -2)$	-3	$-18;\ f(x) < 0$
$(-2, 0)$	-1	$0;\ f(x) > 0$
$(0, 3)$	1	$-6;\ f(x) < 0$
$(3, \infty)$	4	$24;\ f(x) > 0$

What does all of this mean? Look at the graph of $f(x) = x^3 - x^2 - 6x$ shown in the figure on the preceding page. Note that (as shown in the table) $f(x)$ is negative within the intervals $(-\infty, -2)$ and $(0, 3)$ and is positive in the other two intervals. Notice that the range of $f(x)$ is $(-\infty, \infty)$ so its maximum and minimum are undefined. However, there are two *local extrema*, as shown on the graph (Points A and B). Now imagine a line tangent to $f(x)$ at a given point. Clearly at the left side of the curve such a line would have a positive slope, and as it moves along the curve, the tangent slope decreases until the line reaches Point A where its slope is zero, the *local maximum*.

Continuing, the slope of the tangent becomes increasingly negative until it reaches the *inflection point* where its slope begins to increase (become less negative) to zero at Point B, the *local minimum* when the slope again becomes positive. But wait, there's more! Select any two points on $f(x)$; if a line drawn between those points lies above the curve, then the curve between those points is *convex*, or *concave upward*; if the line falls below the curve, then the curve is *concave downward* between the points.[*] At the inflection point the curve changes from convex to concave or vice-versa. Local extrema and inflection points for continuous functions are easy to find through the use of differential calculus.

Arrow notation

Phrases like "as x approaches negative infinity from the right" (*i.e.*, moving to the left towards negative infinity), or as x approaches 5 from the left" (*i.e.*, x is close to and approaching 5 on the left (where $x < 5$) are more easily express using *arrow notation*:[**]

"$x \to -\infty$" and "$x \to 5^-$." •

* The terms "convex downward" and "convex upward" are sometimes used for "concave upward" and concave downward."

** There are other types of "arrow notation" in math.

THE MATH TUTOR'S FRIEND
factors
rational function
repeated factors
It's easy !!
decomposition
Mr. Math
expansion
PARTIAL FRACTION EXPANSION

6. PARTIAL FRACTION EXPANSION

TABLE OF CONTENTS

page

6. PARTIAL FRACTION EXPANSION

Introduction to Partial Fraction Expansion

Also called *Partial Fraction Decomposition*, partial fraction expansion is simply the reverse of finding a common denominator for a sum of fractions. For example, let's look at the sum

$$1/3 + 3/5 - 2/7 \qquad (6\text{-}1)$$

To add these guys together we've got to first find a common denominator. Take the cross-product of the two left terms:

$$(5 + 9)/15 - 2/7 = 14/15 - 2/7$$

Now take the cross-product of the two remaining terms:

$$[(7\times14) - (2\times15)]/(7\times15) = (98 - 30)/105 = 68/105$$

This is old stuff, finding a common denominator; in this case it is also the *least common denominator (LCD)*.

So what? Well, suppose you are given the fraction 68/105 and asked to decompose it into a sum of fractions. How do you do that?

You must answer two questions in order to solve this:

1) How many terms (*i.e.*, how many fractions) are there; and
2) What are the fractions?

To answer the first question, factor the denominator, 105, completely. This is easily done using the divisibility rules: the digits of 105 add to 6, which is divisible by 3, so 105 is divisible by 3

$$\begin{array}{r|l} 3 & \underline{105} \\ 5 & \underline{35} \\ & 7 \end{array}$$

Dividing by 3 yields 35 which we know is divisible by 5 yielding the prime number 7. The denominators are 3, 5, and 7, so there will be three fractions. We don't know the numerators yet so let's call them A, B, and C, respectively (all non-zero):

$$68/105 = A/3 + B/5 + C/7 \qquad (6\text{-}2)$$

We must find the numerators A, B, and C.

Let's clear the denominator in Equation (6-2) by multiplying through by 105:

$$68\times(105/105) = 105A/3 + 105B/5 + 105C/7$$

$$68 = 35A + 21B + 15C \qquad (6\text{-}3)$$

Equation (6-3) has three unknowns; it doesn't look too promising to be able to get a solution. However, observe that

a) A, B, and C must be non-zero, and

b) A = B = C = 1 is not possible because then the sum of the three coefficients on the right of Equation (6-3) is greater than 68, so one or two of the variables must be negative.

Perhaps we can analyze Equation (6-3) by use of a table:

A	B	C	35A+21B+15C	$\stackrel{?}{=}$ 68
1	1	-1	41	no
1	-1	1	29	no
-1	1	1	1	no
1	2	-1	62	no (close)
1	3	-1	83	no (too high)
1	3	-2	68	YES!

$$\frac{68}{105} = \frac{1}{3} + \frac{3}{5} - \frac{2}{7}$$

This is definitely a trial-and-error approach but didn't take too much time (and we do know the answer). This example points out that partial fraction expansion is really the reverse of the LCD computation.

Partial fraction expansion of a rational function

The approach we used applies also to the partial fractions of a rational function whose denominator is factorable.

Let $f(x) = G(x)/H(x)$, where $H(x) \neq 0$ *and* is factorable.

In addition, the degree of the denominator, $H(x)$, must be less than the degree of the numerator, $G(x)$. If not, then you must employ long division or synthetic division.

If these conditions are satisfied, partial fraction expansion of $f(x)$ begins by factoring $H(x)$ completely. That is,

$$H(x) = h_1(x) \times h_2(x) \times h_3(x) \times \ldots \text{ so that}$$

$$f(x) = G(x)/H(x) = A/h_1(x) + B/h_2(x) + C/h_3(x) + \ldots$$

and our task is to determine the unknowns A, B, C, ...

There are three cases to consider:

Case 1 - $H(x)$ has no repeating factors, e.g., x^2, $(x-3)^2$, *etc.*

Case 2 - $H(x)$ has repeating factors, but no non-linear factor that itself is not factorable (e.g., no factor like $2x^2+1$); and

Case 3 - $H(x)$ has at least one non-factorable non-linear factor. (e.g., $2x^2+1$)

Before proceeding, observe that when you equate two expressions, e.g.,

$$ax^2 + bx + c = mx^2 + nx + g$$

the coefficients of x^2 and of x as well as the constants on both sides of the equation must be equal. That is,

$$a = m,\ b = n, \text{ and } c = g.$$

This is called *equating coefficients* and relies on the fact that if two polynomials are identical then the coefficients of the *like* terms are identical.

Partial fraction expansion examples

Case 1 - $H(x)$ has distinct linear factors:

$$G(x) = (x + 1)$$
$$H(x) = x(x + 2)$$
$$f(x) = G(x)/H(x) = (x + 1)/[x(x + 2)] \qquad (6\text{-}4)$$

The factors of $H(x)$ are x and $(x + 2)$ and we write

$$(x + 1)/x(x + 2) = \mathrm{A}/x + \mathrm{B}/(x + 2) \qquad (6\text{-}5)$$

Multiply both sides of Equation (6-5) by $x(x+2)$ to get

$$(x + 1) = \mathrm{A}\,(x + 2) + \mathrm{B}x \qquad (6\text{-}6)$$

Now select a value of x to make one of the terms on the right-hand side equal to zero; let's choose $x = 0$ to get

$$(0 + 1) = \mathrm{A}(0 + 2) + \mathrm{B}\times 0$$
$$1 = 2\mathrm{A}$$
$$\text{from which A} = 1/2$$

Substitute 1/2 for A in Equation (6-6) to get

$$(x + 1) = (x + 2)/2 + \mathrm{B}x \qquad (6\text{-}7)$$

Now select $x = -2$ to isolate B in eEquation (6-7):

$$(-2 + 1) = (-2 + 2)/2 + \mathrm{B}(-2)$$
$$-1 = -2\mathrm{B}$$

from which B = 1/2.

Substituting the values for A and B in Equation (6-5) yields

$$(x + 1)/[x(x + 2)] = 1/(2x) + 1/[2(x + 2)] \qquad (6\text{-}8)$$

which is easily verified by cross-multiplying the right side of Equation (6-8).

Case 2 - $H(x)$ has repeated factors:

$$G(x) = (x + 1)$$
$$H(x) = x^2(x + 2)$$
$$f(x) = G(x)/H(x) = (x + 1)/[x^2(x + 2)] \qquad (6\text{-}9)$$

The factors of $H(x)$ are x, x^2 and $(x + 2)$ and we write

$$(x + 1)/x^2(x + 2) = \mathrm{A}/x + \mathrm{B}/x^2 + \mathrm{C}/(x + 2) \qquad (6\text{-}10)$$

As in Case 1 we multiply through by $H(x)$ to get

$$(x + 1) = \mathrm{A}\,x(x + 2) + \mathrm{B}(x + 2) + \mathrm{C}\,x^2 \qquad (6\text{-}11)$$

Expand the right-hand side of Rquation (6-11):

$$x + 1 = \mathrm{A}x^2 + 2\mathrm{A}x + \mathrm{B}x + 2\mathrm{B} + \mathrm{C}\,x^2$$

Collect like terms

$$x + 1 = (A + C)x^2 + (2A + B)x + 2B$$

Now, equate coefficients of like terms on both side of the equation (if there's no like term then the coefficient is 0):

Equate coefficients of x^2: $0 = A + C$ (a)

from which $C = -A$

Equate coefficients of x: $1 = 2A + B$ (b)

Equate the constants: $1 = 2B$

from which $B = 1/2$

and so (b) above becomes: $1 = 2A + B = 2A + 1/2$

from which $A = 1/4$

and from (a): $C = -A = -1/4$

so, $A = 1/4$; $B = 1/2$; $C = -1/4$

and Equation (6-10) becomes

$$(x + 1)/[x^2(x + 2)] = A/x + B/x^2 + C/(x + 2)$$

$$= 1/(4x) + 1/(2x^2) - 1/[4(x + 2)] \qquad (6\text{-}12)$$

which is verified algebraically by adding the rational functions on the right side of Equation (6-12) to get

$$(x + 1)/x^2(x + 2).$$

Case 3 - $H(x)$ has at least one non-factorable factor:

$$G(x) = (x + 2)$$

$$H(x) = (x+1)(x^2 + x + 1)$$

$$f(x) = G(x)/H(x) = (x + 2)/[(x + 1)(x^2 + x + 1)] \qquad (6\text{-}13)$$

Here, $(x^2 + x + 1)$ is a quadratic with no real solution. It is possible to factor this (using the quadratic formula) to get the conjugate pair of complex numbers $x + (-1 + i\sqrt{3})/2$ and $x + (-1 - i\sqrt{3})/2$. However, we are dealing here only with real numbers and so consider the quadratic $x^2 + x + 1$ as non-factorable.

Expanding Equation (6-13):

$$(x+2)/[(x+1)(x^2+x+1)] = A/(x+1) + (Bx+C)/(x^2+x+1) \qquad (6\text{-}14)$$

Because its denominator is non-factorable, the last term in Equation (6-14) must have the linear expression $(Bx + C)$ for its numerator rather than a single constant as in Case 1 and Case 2. Proceeding as before we multiply through by the denominator on the left-hand side:

$$(x + 2) = A(x^2 + x + 1) + (Bx + C)(x + 1)$$

Next, expand the right-hand side

$$x + 2 = Ax^2 + Ax + A + Bx^2 + Bx + Cx + C$$

and then collect like terms:

$$x + 2 = (A + B)x^2 + (A + B + C)x + (A + C)$$

Now, equate the coefficients of like terms

equate coefficients of x^2: $0 = A + B$ from which $B = -A$ (a)

equate coefficients of x: $1 = A + B + C$ (b)

equate constants: $2 = A + C$ from which $C = 2 - A$ (c)

and so from (b) and (c): $1 = B + 2$ from which $B = -1$

and from (a): $A = -B = -(-1) = 1$

from (b): $C = 1 - A - B = 1 - 1 - (-1) = 1$

Equation (6-14) becomes

$$(x + 2)/[(x + 1)(x^2 + x + 1)] = 1/(x + 1) + (1 - x)/(x^2 + x + 1)$$

which is easily verified.

Application of partial fraction expansion

Partial fraction expansion is especially useful in calculus for solving differential and integral equations algebraically using *LaPlace Transforms*.

•

INDEX

Symbols

A

B

C

D

E

F

R

S

T

U

V

W

X

Z

• • •

TEST QUESTIONS

ALGEBRA II

QUIZZES

• • •

• Let $f(x) = \sqrt{(x-3)}$ (i.e., the square root of $(x-3)$. The domain of this function is

A. All x such that $x > 0$

B. All x such that $x < 0$

C. The interval $[0, \infty)$

D. The interval $(-\infty, 3)$

E. None of the above

• • Answer E. None of the above.

The domain of a function consists of all allowable values of x. Here, we have a square root function and the square root of negative numbers are not allowed. To find the domain, determine the values of x that will make the radicand $(x-3)$ negative. Clearly, $x < 3$ will make the radicand negative, so the domain of $\sqrt{(x-3)}$ is the interval $[3, \infty)$, that is, the domain is all x such that x is equal to or greater than 3.

• ♥ •

• A quadratic equation is

A. always a trinomial of the form $ax^2 + bx + c = 0$

B. may have the binomial form $ax^2 + bx = 0$

C. may have the monomial form $ax^2 = 0$

D. may be any of the above

E. may be "A" or "B" above only

• • Answer D.

The second form (B) can be expressed as the trinomial $ax^2 + bx + 0 = 0$. The third form (C) can be expressed as the trinomial $ax^2 + 0x + 0 = 0$

• Solve for x:

$$|x| = x^2$$

A. $x = 1$; $x = -1$; $x = 0$

B. $x = 2$; $x = -2$

C. $x = 3$; $x = -3$; $x = 0$

D. All of the above

E. No real solution

• • Answer A. Just plug in the values. You can apply the definitions:
Either $x^2 - x = 0$, yielding $x(x - 1) = 0$ and $x = 0,1$
or $x^2 + x = 0$, yielding $x(x + 1) = 0$ and $x = 0, -1$

• ♥ •

• Which is a correct step when factoring by grouping?

$$2x^2 + 16x + 25 = -7$$

A. $(x + 4)(x + 4) = 0$

B. $x^2 + 8x + 16 = 0$

C. $x^2 + 4x + 4x + 16 = 0$

D. All of the above

E. None of the above

• • Answer D. First, add 7 to both sides of the equation to get B, then split $8x$ to get C, then factor $(x + 4)$ to get A. Note that answer B is a perfect square.

• Express $|x-4|<7$ as a compound inequality.

A. $x = 3$

B. $-3 < x < 11$

C. $-7 < x < 7$

D. $-4 < x < 4$

E. None of the above

• • Answer B.

$\|x-4\|<7$	[original inequality]
$-7 < x-4 < 7$	[definition of absolute value]
$-7+4 < x-4+4 < 7+4$	[add 4 to all sides]
$-3 < x < 11$	[the answer]

• ♥ •

• What are the zeroes for

$$f(x) = x^3 - x^2 - 6x\ ?$$

A. 1, 1, -6
B. 1, 1, 6
C. -2, 0, 3
D. -1, 0, 1
E. None of the above

• • Answer C.

$x^3 - x^2 - 6x$	[original]
$x(x^2 - x - 6)$	[factor out x]
$x(x-3)(x+2)$	[factor the quadratic]

The three zeroes are $x = -2$, $x = 0$, $x = 3$

• Complete Pascal's Triangle to find the coefficients of expanded $(a+b)^7$

1
1 1
1 2 1
1 3 3 1
1 4 6 4 1

A. 1, 4, 6, 4, 1

B. 1, 5, 10, 10, 5, 1

C. 1, 6, 15, 20, 15, 6, 1

D. 1, 7, 21, 35, 35, 21, 7, 1

E. None of the above

• • Answer D. Append three rows to get D. $(a+b)^7 =$

$$a^7b^0 + 7a^6b^1 + 21a^5b^2 + 35a^4b^3 + 35a^3b^4 + 21a^2b^5 + 7a^1b^6 + a^0b^7$$

Note that each term is of degree 7 (*i.e.*, variable exponents add to 7)

$$= a^7 + 7a^6b + 21a^5b^2 + 35a^4b^3 + 35a^3b^4 + 21a^2b^5 + 7ab^6 + b^7$$

• ♥ •

• For the quadratic equation of the form $ax^2 + bx + c = 0$, the expression "$b^2 - 4ac$" in the related quadratic formula is called the

A. "discriminant"

B. "quadratic factor"

C. "middle term"

D. "quadrant"

E. None of the above

• • Answer A.

• Find the GCF of

2490, 2150, and 5212.

A. 2

B. 4

C. 5

D. 43

E. There is no common factor.

• • Answer A. The factors of 2490 are 2, 3, 5, 83

of 2150 are 2, 5^2, 43

of 5212 are 2^2, 1303

The greatest factor common to all three numbers is 2.

• ♥ •

• The greatest common factor of f(x) is ______ .

$$f(x) = 2x^5 - 6x^4 - 2x^3 + 6x^2$$

A. $2x^5$

B. $6x^2$

C. $2x^2$

D. $2x^3$

E. None of the above

• • Answer C. $2x^2$ is the greatest term common to all of the terms in f(x). You can solve this by inspection.

• Solve the system of equations for x:

$$y = |\,2x + 3\,|$$
$$y = 5$$

A. 0, 5

B. 2, 3, 5

C. -3, 5

D. -4, 1

E. None of the above

• • Answer D.

The two equations are $2x + 3 = 5$ and $2x + 3 = -5$ (or $-2x - 3 = 5$)

• ♥ •

• The parent function of $y = 17(x - 4)^3 - 34$ is

A. 17

B. 34

C. x^3

D. $(x - 4)^3$

E. None of the above

• • Answer C.

- $-(-|-3|^{2})^{3} =$

  A. -9

  B. -729

  C. 729

  D. -81

  E. None of the above

• • Answer C.

$$|-3|^{2} = 9 \text{ so that } -(-|-3|^{2}))^{3} = -(-9)^{3} = -(-729) = 729$$

Note that there are an even number (2) of negative signs (the negative sign withing the absolute-value bars does not affect the results); therefore, the product is positive.

• ♥ •

- What is the meaning of | 4 | in terms of the number line?

  A. The distance to 0 is 8

  B. The distance to 0 is 4

  C. Can't express an | 4 | on the number line

  D. The distance to 0 is 0

  E. None of the above

• • Answer B.

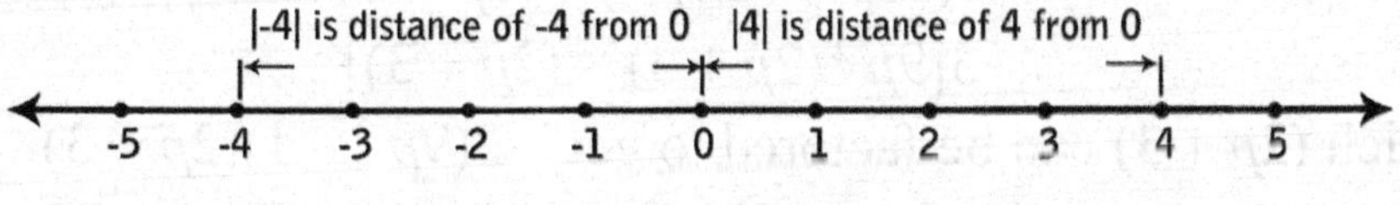

• Factor completely: $8x^3 - 64x^2 + x - 8$

[Hint: factor by grouping]

A. $(8x^2 + 1)(x - 8)$

B. $(x - 8)^3$

C. $(8x^2 + 1)^2$

D. All of the above

E. None of the above

• • Answer: A. Factor by grouping. First, factor $8x^2$ from the first two terms:

$$8x^2(x - 8) + (x - 8),$$

now we factor $(x - 8)$ from these two terms to get the answer:

$$(8x^2 + 1)(x - 8)$$

• ♥ •

• Factor completely:

$$54p^3 + 81p^2 - 6p - 9$$

A. $(2p + 3)^3$

B. $3(3p + 1)(3p - 1)(2p + 3)$

C. $(3p + 1)(3p - 1)^2$

D. $3(18p^3 + 27p^2)$

E. None of the above

• • Answer B.

First note that 3 can be factored from each term: $3(18p^3 + 27p^2 - 2p - 3)$
Now proceed to factor by grouping

$$3(18p^3 + 27p^2) - (2p + 3)$$
$$3[9p^2(2p + 3) - (2p + 3)]$$

from which $(2p + 3)$ can be factored to get $3(9p^2 - 1)(2p + 3)$
that contains the factor $(9p^2 - 1)$ which we recognize as the difference of two squares, so proceeding... $3(3p + 1)(3p - 1)(2p + 3)$

• The GCF and LCM of 60, 12, $6x$ and $30x^2$ are, respectively,

A. 2 and $3x^2$
B. 6 and $60x^2$
C. 30 and $30x$
D. 2 and x
E. None of the above

• • Answer B. Set up the table as follows:

	Factors	*Product*
60:	2 2 3 5	
12:	2 2 3	
6x:	2 3 x	
$30x^2$:	2 3 5 x x	
GCF:	2 3	6
LCM:	2 2 3 5 x x	$60x^2$

• ♥ •

• Factor 3880 completely into its prime factors.

{ ________________________________ }

• • Answer {2, 2, 2, 5, 97} or preferably $\{2^3, 5, 97\}$

2 | 3880
2 | 1940
2 | 970
5 | 485
97

• The critical points for the absolute-value equation shown below are:

$$| x + 1 | = | -5x + 8 |$$

A. 1 and 8
B. -1/5 and -8
C. -1 and 8/5
D. 1 and -8/5
E. None of the above.

• • Answer C.
The critical points are the values of x that set the expressions between the "| |" signs to zero.

• ♥ •

• Solve $| 3x - 2 | + 4 = 3$ for x

A. $x = 7$
B. $x = -1$
C. $x = 1$
D. $x = 1/3$
E. None of the above

• • Answer E. There is no solution.
Subtract 4 from both sides of the " = " sign yields $| 3x - 2 | = -1$ which is not possible (absolute values cannot be negative).

• Solve $| 2x - 5 | = 1$ for x. Do not include extraneous solutions, if any.

A. 2, 1

B. 2

C. 3, 2

D. 3

E. None of the above

• • Answer C.

Either ***(a)*** $2x - 5 > 0$ or ***(b)*** $2x - 5 < 0$.

(b) $2x - 5 > 0$: remove the absolute-value bars and solve for x: $2x = 6$ and $x = 3$.

(c) $2x - 5 < 0$: $-2x + 5 = 1$ $-2x = -4$ $x = 2$

(make one side of the equation the negative of the original; here, change $2x - 5$ to $-2x + 5$). Note: For ***(b)*** we could have used "$2x - 5 = -1$," then $2x = -1 + 5$ to get the same result, i.e., $x = 2$.

Thus, $x = 3$ or $x = 2$.

Now confirm these values in the original equation, $| 2x - 5 | = 1$.

$$x = 3: | 2\times3 - 5 | = | 6 - 5 | = | 1 | = 1 \checkmark$$
$$x = 2: | 2\cdot2 - 5 | = | 4 - 5 | = | -1 | = 1 \checkmark$$

Therefore, both solutions are valid (i.e., not extraneous).

• ♥ •

• $(-|-7|)^2 =$

A. 49

B. -49

C. $|7|$

D. $-|7|$

E. None of the above

• • Answer A.

Applying the order of operations, $|-7| = 7$, and the contents of the parentheses becomes $(-7)^2 = 49$.

• The greatest common factor (GCF) of the integers 2940 and 3150 is

A. 249

B. 350

C. 210

D. 315

E. None of the above

• • Answer C.

	Factors	*Product*
2940:	2 2 3 5 7 7	
3150:	2 3 3 5 5 7	
GCF:	2 3 5 7	$2 \times 3 \times 5 \times 7 = 210$
LCM	2 2 3 3 5 5 7 7	$2 \times 2 \times 3 \times 3 \times 5 \times 5 \times 7 \times 7 = 44{,}100$

• ♥ •

• The least common multiple (LCM) of 2940 and 3150 is

A. 44,100

B. 88,299

C. 210

D. 5640

E. None of the above

• • Answer A. (See answer to previous question)

• Find the quadratic with a zero at $x = \sqrt{11}$ that passes through the point (-1, 20)

A. $f(x) = -2(x^2 - 1) = -x^2 + 2$

B. $f(x) = -2(x^2 - 1) = -2x^2 + 22$

C. $f(x) = (x^2 - 11) = -2x^2 + 22$

D. $f(x) = -2(x^2 - 11) = -2x^2 + 22$

E. None of the above.

• • Answer D. The roots are the conjugates $x=\sqrt{11}$ and $x = -\sqrt{11}$. We don't know yet what the coefficient of x^2 is, so we call it a in our equation:

$$y = a(x - \sqrt{11})(x + \sqrt{11}) = a(x^2 - 11) \text{ [difference of squares]}$$

Since this curve passes through the point (-1, 20), we substitute 20 for y and -1 for x, then write $20 = a[(-1)^2 - 11] = a(-10)$. Solving for a, we get $a = 20/(-10) = -2$, so the answer is

$$y = f(x) = -2(x^2 - 11) = -2x^2 + 22.$$

Note that there is no "x" term (its coefficient is 0), but we still have a quadratic.

• ♥ •

• $(3x^3 - 2x + 5) \div (x - 1) =$

A. $3x^2 + 3x + 1$ with no remainder

B. $3x^2 + 3x + 1$ with remainder 6

C. $3x^2 + 3x + 1$ with remainder 1

D. $3x^2 + 3x + 1$ with remainder 3

E. None of the above

• • Answer B. Use long division or synthetic division (the easier method).

• In order to factor $x^2 + 7x + 10$ by grouping we would first

A. Add $(-7x)$ to both sides of the equation

B. Split the middle term into $x + 6x$

C. Split the middle term into $2x + 5x$

D. Split the middle term into $3x + 4x$

• • Answer C.

Note that among these, the only factor of the quadratic constant 10 is 5, so we select $2x + 5x$ and rewrite the equation as $x^2 + 2x + 5x + 10$. Look at the first two terms. They factor into $x(x + 2)$. The last two terms factor into $5(x + 2)$, so we have a common factor $(x + 2)$:

$$x^2 + 7x + 10 = x(x + 2) + 5(x + 2) = (x + 2)(x + 5)$$

• ♥ •

• $(a + b)^0 =$

A. 0

B. $a\ b$

C. $(a + b)$

D. 1

E. None of the above

• • Answer D. Anything (except zero) raised to the zero power equals one.

x
y z
1 2 1
1 3 3 1
1 4 6 4 1

The values of the top two lines of Pascal's Triangle (shown as x, y, and z above) are

A. $x = 0$, $y = 0$, and $z = 0$

B. $x = 0$, $y = 0$, and $z = 1$

C. $x = 0$, $y = 1$, and $z = 1$

D. $x = 1$, $y = 1$, and $z = 1$

E. None of the above

• • Answer D. Each number in a line is the sum of the two numbers directly above it. If the "missing" numbers to the left (and right) of the 1s are 0.

1
1 1
1 2 1
1 3 3 1
1 4 6 4 1

• Simplify the following:

$$(a^4 - b^4)$$

A. $(a + b)(a - b)$

B. $(a^3 - b^3)(a^2 - b^2)(a - b)$

C. $(a^3 + b^3)(a^2 \quad b^2)$

D. $(a^2 + b^2)(a + b)(a - b)$

E. None of the above

• • Answer D.

This is a *difference of squares* pattern.

$$(a^4 - b^4) = (a^2 + b^2)(a^2 - b^2) = (a^2 + b^2)(a + b)(a - b)$$

• Factor $2x^4$ from each term of the following expression:

$$2x^4 - 4x^3 + 8x^2 + 2x.$$

A. $2x^4(1 - 4x^3 + 8x^2 + 2x)$

B. $2x^4(1 - 2x^{-1} + 4x^{-2} + x^{-3})$

C. $2x^4(2x^4 - 4x^3 + 8x^2 + 2x)$

D. All of the above

E. The expression is not factorable

• • Answer B. Verify this by applying the Distributive Law:

$2x^4(1 - 2x^{-1} + 4x^{-2} + x^{-3})$

$= (2x^4 - 2x^4 \times 2x^{-1} + 4x^{-2} \times 2x^4 + x^{-3} \times 2x^4)$

Simplify by use of the rules for exponents

$= 2x^4 - 4x^3 + 8x^2 + 2x$, the original expression.

• ♥ •

• Factor 1,890 completely

A. 2, 3, 3, 3, 5, 7

B. 6, 9, 5, 7

C. 5

D. 2, 3, 5, 7

E. None of the above

• • Answer A.

Preferably written as 2, 3^3, 5, 7

• The graph shown to the right is an example of a

A. Continuous function

B. Linear function

C. Piecewise function

D. Quadratic function

E. None of the above

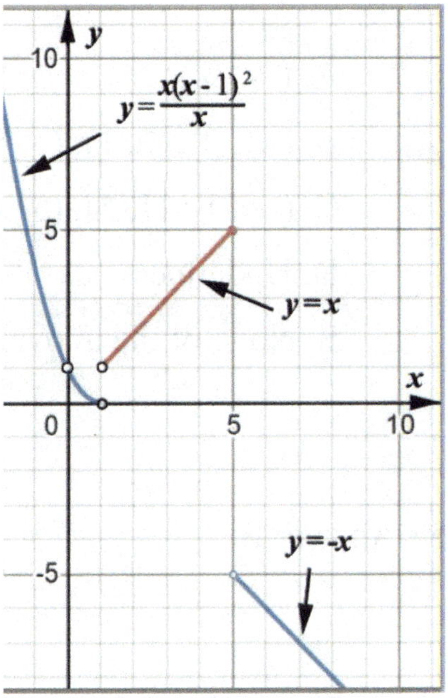

• • Answer C.

• ♥ •

• The piecewise function shown in the graph has

A. a "hole" at $x = 0$

B. two "holes" at $x = 1$

B. "holes" at $x = 1$ and $x = 5$

D. a vertical asymptote at $x = 1$

E. None of the above

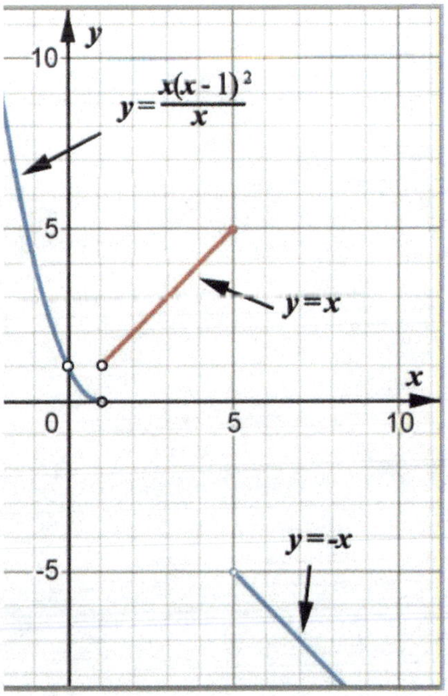

• • Answer A. The function contains the fraction x/x which is undefined at $x = 0$.

• The domain of the piecewise function shown here is

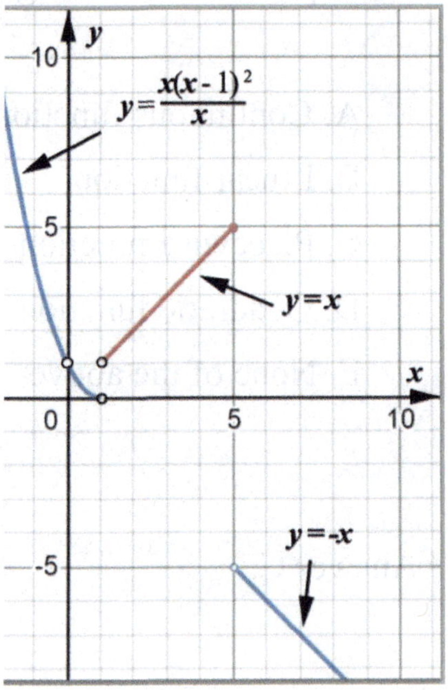

A. $(-\infty, \infty)$

B. $(-\infty, 0), (0, 1), (1, 5], [5, \infty)$

C. $(-\infty, 1), (1, 5], (5, \infty)$

D. Undefined

E. None of the above

• • Answer B.

• ♥ •

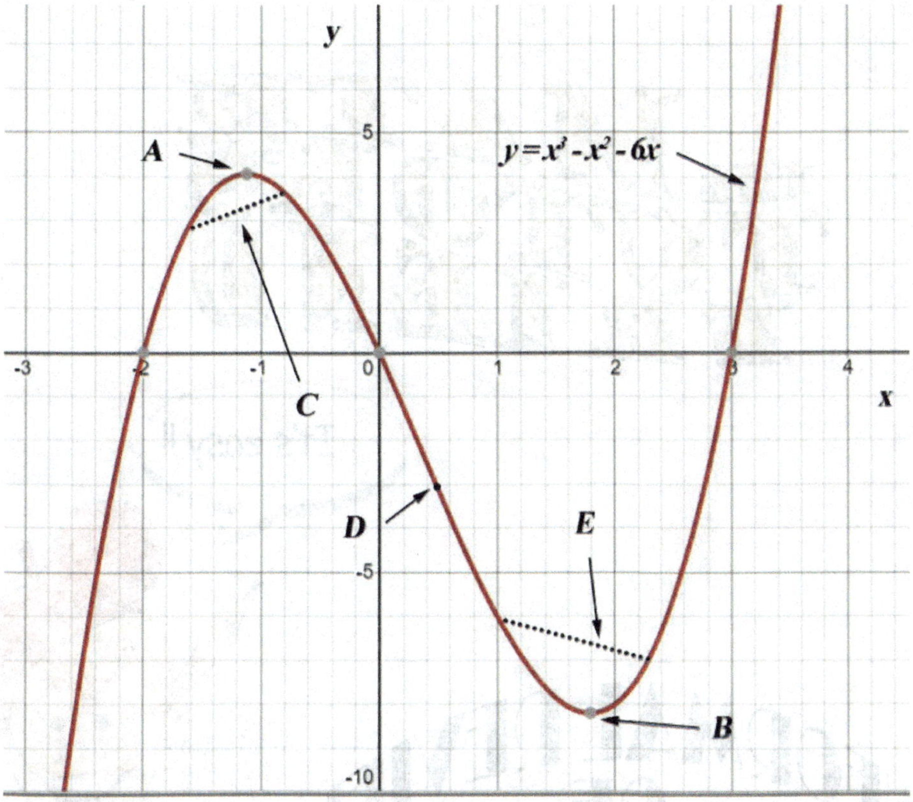

• Identify the labels in the above graph.

A -
B -
C -
D -
E -

1. Local minimum
2. "Hole"
3. Convexity
4. Concavity
5. Local maximum
6. Global minimum
7. Global maximum
8. Inflection point

A _____; B _____; C _____; D _____; E _____

• • Answer;

A - 5; B - 1; C - 3; D. - 8; E - 4

C is convex because the line segment lies below the curve.

E is concave because the line segment lies above the curve.

THE MATH TUTOR'S FRIEND

INFORMAL REVIEW OF HIGH-SCHOOL MATH TOPICS FOR TUTORS

-3-

' Algebra II '

Made in the USA
Monee, IL
02 November 2026

67386700R00077